Maladies of Modernity

Other Books of Interest from St. Augustine's Press

Ellis Sandoz, *The Politics of Truth and Other Untimely Essays:*
The Crisis of Civic Consciousness

Ellis Sandoz, *Give Me Liberty:*
Studies in Constitutionalism and Philosophy

Barry Cooper, *Consciousness and Politics*

Montgomery Erfourth, *A Guide to Eric Voegelin's Political Reality*

Gerhart Niemeyer, *Between Nothingness and Paradise*

Josef Siefert, *Christian Philosophy and Free Will*

Josef Siefert, *True Love*

Alexandre Kojève, *The Concept, Time, and Discourse*

Peter Kreeft, *Summa Philosophica*

Peter Kreeft, *Socrates' Children: The 100 Greatest Philosophers*

Peter Kreeft, *Ecumenical Jihad*

Zbigniew Janowski, *Augustinian-Cartesian Index*

Jean-Luc Marion, *Descartes's Grey Ontology:*
Cartesian Science and Aristotelian Thought in the Regulae

Daniel J. Mahoney, *The Other Solzhenitsyn:*
Telling the Truth about a Misunderstood Writer and Thinker

Philippe Bénéton, *The Kingdom Suffereth Violence: The Machiavelli /*
Erasmus / More Correspondence and Other Unpublished Documents

Peter Lawler, *Allergic to Crazy: Quick Thoughts on Politics,*
Education, and Culture, Rightly Understood

Josef Kluetgen, s.j., *Pre-Modern Philosophy Defended*

James V. Schall, *The Sum Total of Human Happiness*

Josef Pieper, *Enthusiasm and Divine Madness*

Josef Pieper, *Don't Worry about Socrates*

Roger Scruton, *The Politics of Culture and Other Essays*

Roger Scruton, *An Intelligent Person's Guide to Modern Culture*

Maladies of Modernity

Scientism and the Deformation of Political Order

David N. Whitney

ST. AUGUSTINE'S PRESS
South Bend, Indiana

Manufactured in the United States of America.

1 2 3 4 5 6   25 24 23 22 21 20 19

Library of Congress Cataloging in Publication Data
Whitney, David N.
Maladies of modernity: scientism and the deformation of political order
David N. Whitney.
pages cm
ISBN 978-1-58731-489-6 (paperback)
1. Science – Social aspects. 2. Science – political aspects.
3. Science –Philosophy. 4. Scientism. I. Title.
Q175.5.W49 2014
303.48'3 – dc23   2014009624

∞ The paper used in this publication meets the minimum requirements of the American National Standard for Information Sciences - Permanence of Paper for Printed Materials, ANSI Z39.48-1984.

St. Augustine's Press
www.staugustine.net

Table of Contents

Acknowledgments

I would like to thank Ellis Sandoz for the invaluable insight and guidance that he has provided over the past decade. He graciously served as my graduate advisor at Louisiana State University and showed a genuine concern for my intellectual and personal development. I would also like to thank all of those who provided feedback on conference papers and my dissertation since I drew heavily from both while composing this manuscript. And finally, I must thank my friends and family, and especially my parents, for their unwavering support.

Chapter 1
Scientism and the Dogmatics of Modernity

"The damage of scientism is done. . . . [T]he insane have succeeded in locking the sane in the asylum. From this asylum no physical escape is possible. As a consequence of the interlocking science and social power, the political tentacles of scientistic civilization reach into every nook and corner of an industrialized society, and with increasing effectiveness stretch over the whole globe. There exist only differences, though very important ones, in the various regions of the global asylum with regard to the possibility of personal escape into the freedom of the spirit. What is left is hope—but hope should not obscure the realistic insight that we who are living today shall never experience freedom of spirit in society."[1]—Eric Voegelin

During his first Inaugural Address, President Barack Obama vowed to "restore science to its rightful place."[2] This certainly did not come as a surprise to anyone who even casually followed his campaign. The issues of climate change and embryonic stem cell research came up in numerous speeches and debates throughout Obama's presidential campaign. The foundation had been set several years earlier as former Vice President Al Gore successfully brought the issue of global warming to the forefront of political debate, primarily through his book and subsequent documentary, *An Inconvenient Truth*. And record high oil prices in the midst of a receding economy all but ensured that the topic would be of utmost concern in the election season. Not even halfway into his first term, President Obama made good on his campaign promises to push for environmental regulations as well as to aggressively promote embryonic stem cell research. Through these and subsequent actions, President Obama could claim to be making good on his promise to "restore science to its rightful place." But considering the stakes, we owe it to ourselves to seriously consider what science's "rightful place" truly is within society. The answer to that question carries serious political consequences and thus deserves our full attention.

The interplay between politics and science is certainly not a new phenomenon, but the unprecedented scientific advances in medicine and technology during the twentieth century, and into the first decade of the twenty-first, have forced us to reconsider that relationship. While advances in medicine and technology led to unprecedented wealth and accumulation of material comforts, man's increased control over nature also meant that he could destroy his fellow man with greater ease. Thus, in addition to penicillin, vaccinations for crippling diseases such as polio and smallpox, the mass production of the automobile, television, the internet and countless other beneficial inventions, the twentieth century saw tens of millions die at the hands of their own governments.[3] Atomic energy could provide power for a whole city, but it could also be used to decimate that same city. Therefore, a series of questions must be raised as to the proper relationship between science and politics. Should governments control scientific research agendas? Should they determine what inventions to allow? And if so, how can such determinations be made? Who ultimately decides if a scientific advancement is good or bad for society? The answers to such questions go a long way in determining the role of science within a society. What cannot be disputed in contemporary politics is that science carries great weight. No matter what the issue may be (stem cell research, climate change, militarized weapons, etc.), the importance of having science on "your side" cannot be overestimated. In many cases, both sides will claim to have scientific support for their arguments (this certainly has been the case in the climate change debates). The reason for this is simple: science has produced countless useful innovations, especially within the past century. Nobody has to argue for the power of science or its utility to society, as the evidence is overwhelming in everyday life. However, that prestige can be abused and exploited to advance political agendas. One need look no further than the Nazis' "scientific" race theories during WWII to see the horrendous results of such an abuse.[4] Under the auspices of "science," the Nazis were able to advance their horrific eugenics programs. According to this argument, the Aryan race was biologically superior and thus had license to dominate and destroy the sub-human races. Of course, the "science" used to come to this conclusion was not scientific at all, but rather a politically motivated pseudo-science. Nevertheless, the claim of science served to legitimate the political actions that followed.

Much like with religion, the abuses committed in the name of science should not condemn the whole enterprise. Therefore, it is important to distinguish between genuine science and pseudo-science. Science in the broadest sense is the search for the truth about the nature of things. Modern science is essentially synonymous with natural science, or what used to be referred to as natural philosophy, and is characterized by the scientific method. Experimentation is the hallmark of modern science as nature is "vexed" in order to understand its workings. And to say that the scientific method has led to positive results within the realm of natural science would be to understate the case. Yet, it is important to note *why* modern science has enjoyed success. It is not due to the inherent worth of the method, but rather because the method was designed with the subject matter in mind. The mistaken belief in the inherent power of the method, due in part to the incredible successes of natural sciences such as physics and biology, led to the adaptation of the method to fields that it was not designed for. This was most apparent in the rise of the social sciences. The majority of the scholarship on scientism focuses on this crucial error. First, I will briefly explore some of the seminal critiques, as my study is indebted to the foundation provided by them.[5]

Past Critiques

The critiques of scientism have generally been directed towards the intellectual manifestations, primarily within philosophy and the academy. Eric Voegelin's *New Science of Politics* serves as one of the most striking examples. Voegelin's call for a new science of politics was spurred by the inadequacy of positive political science to address the political crises of his time, especially the experience of Nazi Germany. According to Voegelin's analysis, a restoration of political science is required because of two particularly destructive effects of positivism.[6] The first is the assumption that the methods of the natural sciences have inherent worth and can be applied to all other sciences, including the human sciences. The second, and more dangerous, assumption is that the methods of the natural sciences become the criterion for theoretical relevance. As Voegelin notes:

> The second assumption is the real source of danger. For this second assumption subordinates theoretical relevance to method and thereby perverts the meaning of science. Science is a search for truth concerning the nature of the various realms of being. Relevant in science is whatever contributes to the success of this search. Facts are relevant in so far as their knowledge contributes to the study of essence, while methods are adequate in so far as they can be effectively used as a means for this end. Different objects require different methods.[7]

Voegelin's critique of positivism is echoed by Leszek Kolakowski in *The Alienation of Reason*. Although Kolakowski's work is intended more as a history of positivist thought than a critique, he does raise several critical questions about positivism. He points out that the more radical forms of positivism impose a language that "exempts us from the duty of speaking up in life's most important conflicts, encases us in a kind of armor of indifference to the *ineffabilia mundi*, the indescribable qualitative data of experience."[8] The value-free language prohibits us from making critical judgments and discourages us to think about the fundamental questions of existence since those are deemed meaningless.[9] Furthermore, "Positivism, when it is radical, renounces the transcendental meaning of truth and reduces logical values to features of biological behavior." It results in "the reduction of all knowledge to biological responses."[10] In other words, man's physical nature is stressed to the expense of his spiritual capacities.

F. A. Hayek's *The Counter-Revolution of Science* stands as perhaps the most comprehensive study of scientism in the twentieth century. Like Voegelin, Hayek points to the problematic influence of the natural sciences on the social sciences. As the successes in the natural sciences became more apparent, those in the social sciences began to take note. They became "increasingly concerned to vindicate their equal status by showing that their methods were the same as those of their brilliantly successful sisters rather than by adapting their methods more and more to their own particular problems. . . . [T]his ambition to imitate Science in its methods rather than its spirit has now dominated social studies."[11] Hayek is quick to point out that the claims to an inherent virtue for the methods of the natural sciences were "mostly advanced by men whose right to speak on behalf of the scientists was not above suspicion."[12] Moreover, "the methods which scientists or men fascinated by the natural sciences have so often tried to force upon the social sciences were not always necessarily those which the scientists in fact followed in their own field, but rather those which they believed that they employed."[13] In other words, the average scientist is not to blame for the rise of the scientistic attitude and those that do propound the dogma often do not even understand how natural science is actually practiced.

Hayek's criticism then is not aimed at science itself and is not intended to raise questions as to its value. Instead, he is concerned about the "slavish imitation of the method and language of Science" and with the "mechanical and uncritical application of habits of thought to fields different from those in which they have formed."[14] The *scientistic* view, unlike the scientific, is a "very prejudiced approach, which, before it has concerned its subject, claims to know what is the most appropriate way of investigating it."[15]

Leo Strauss and Michael Oakeshott, while not offering full treatises on the subject, have offered important contributions. Strauss's "Epilogue" from *Liberalism: Ancient and Modern* reminds us "political science stands or falls by the truth of the *prescientific* awareness of political things."[16] Common sense plays a crucial role in the understanding of political things, yet the narrow empiricism of the new political science, often called behavioralism, "must reduce the political things to nonpolitical data" in its attempted break with our common sense understanding of politics.[17] And as Strauss further notes, that break cannot be completed because "empiricism cannot be established empiricistically: it is not known through sense data that the only possible objects of perception are sense data."[18] Like Voegelin and Hayek, Strauss is concerned with the subordination of theoretical relevance to method. He concedes that the logic of the new science may provide criteria of exactness, but "it does not provide criteria of relevance."[19] Instead, criteria of relevance are inherent in the prescientific knowledge that is excluded from the new science. In other words, informed citizens are able to distinguish between important and unimportant political matters.[20] And without a proper place for prescientific knowledge, political science cannot distinguish between relevant and useless information.

Michael Oakeshott's *Rationalism in Politics* gets at a similar notion. Oakeshott provides a reaction "against the modern quest for a science of social life."[21] He outlines two types of knowledge, practical and technical, and argues that the rationalism of modernity has focused on technical knowledge, due to its popularization by Francis Bacon. This limits what can be studied and fails to realize the importance of practical knowledge. The result is an ignorance of tradition, a belief in the infallibility of reason, and a politics of uniformity.[22] Technical knowledge focuses on a specific way of doing something. It designates a technique and gives rules and formulations on its implementation.[23] On the other hand, practical knowledge cannot be formulated into rules. It is a type of common sense that has its use in actual practice.[24] One can think of this as activity or participation. Oakeshott makes the strong case that these two types of knowledge are inseparable. They need each other to function properly.

The critical flaw in modern rationalism is that it focuses on technical knowledge and excludes practical knowledge from the realm of legitimate knowledge. It regards practical knowledge as unimportant because technical knowledge is the *only* true knowledge.[25] This simply does not reflect reality, however. Politics necessarily includes both types of knowledge. The certainty sought through technical knowledge cannot be obtained in politics. As Aristotle noted over two thousand years ago, politics is an imprecise endeavor, yet the modern rationalist rejects this. Tradition plays a key role in understanding politics, but rationalism disregards this facet of knowledge. It *requires* an intellectual purge so that the technique can be learned. Oakeshott's attempt to restore dignity to practical knowledge can be likened to Strauss's appeal to prescientific knowledge and common sense experience.

Assuming that the political philosopher's primary task is to act as a physician for the soul, credit is due to the aforementioned scholars for fulfilling the role of diagnostics, particularly within the intellectual realm.[26] They all drew attention to an acute problem, which has affected both the academy and society as a whole. However, the therapeutic role has not been adequately fulfilled. On the academic side, few studies have sought to offer a suitable replacement for positivism.[27] Intellectual revolutions, much like political revolutions, must be concerned with what will replace the old structure or regime. Thus, the constructive part of my study will outline the characteristics that a new science of politics must encompass if it is to adequately address the political problems of modernity. This will only be a first step, but considering the consequences of inaction, it is a necessary one.

Scientism is a deformation of science and arrogates the name of science to pseudo-scientific, and often politically motivated, endeavors. It refers to the intellectual movement that places primacy on the methods of the natural sciences. It can be characterized as a pseudo-religion or a form of idolatry since its adherents express a dogmatic faith in the power of science. As Eric Voegelin notes, "Science becomes an idol that will magically cure the evils of existence and transform the nature of man."[28] This ignores the limitations of the scope of science and mistakenly ascribes transformative power to it. Again, it must be stressed that this attitude is *not* common among practicing scientists today.[29] Natural scientists should not receive the blame for the ills of scientism. Instead, one must look to the experiences that have engendered the scientistic attitude and to those who have propagated the scientistic dogma. More specifically, the origins of scientism must be explored. While past studies have acknowledged scientism as a problem, there are only a few full treatments on the subject. And the studies dedicated to it tend to focus on the intellectual and epistemological problems caused by scientism without explicitly addressing the political problems. It is my argument that scientism stands as perhaps the key *political* crisis of our age. I seek to show that scientism is not only intellectually impoverishing, but also politically *dangerous*. Thus, after examining the primary sources of scientism within modernity, I will offer a solution that is designed to vitiate the political *and* intellectual effects of scientism.

In spite of these ambitious aims, I realize that the problem of scientism is as old as modernity itself and has become entrenched in contemporary politics.[30] A satisfactory solution then will not come about through a monograph, or even a series of studies on the matter. Yet, such a study can provide a first step in the therapeutic process that is necessitated by the dire political situation. Before any constructive solution can be attempted however, I must demonstrate, and not simply assume, that scientism is intellectually impoverishing and politically dangerous. The majority of my study is committed to such a demonstration. Throughout the study, I will focus on several important themes that remain constant within the scientistic movement including: the dogmatic faith in the methods of the natural sciences (and the accompanying assumption that those methods can be successfully imported into the social sciences), a materialistic worldview, the rejection of the *bios theoretikos*, the prohibition of philosophical questions, and an emphasis on immanent fulfillment through the power of science.

I begin with Francis Bacon and argue that his project represents the foundation of scientism. Through an analysis of the *Novum Organum*, I show how Bacon viewed his own project as a rejection of ancient and medieval philosophy, and particularly of Plato and

Aristotle. I then note the foundation of his new science; one rooted in what we now refer to as the scientific method. Of particular importance, I demonstrate how Bacon's emphasis on that method shapes his thought and especially his politics. I then explore Bacon's *New Atlantis* as it represents the most coherent and clearest picture of the end, or final cause, of his project. It presents us with a scientific utopia; a society based on the principles of Baconian science. The society is controlled by a group of technically trained scientists who have been able to devise ways, through the help of Baconian science, to overcome the perpetual problems of human existence such as disease, lust, greed, and warfare. More specifically, science has provided for the means to control chance, or vicissitude. The scientists of Solomon's House decide what inventions will be "allowed" into society and they serve in a "parent-like" role. Bacon's vision raises several important issues such as *how* the scientists determine which inventions should be allowed. On what basis can those decisions be made? Such decisions require political judgment, yet Bacon subsumes politics under his overall method; a method designed for natural philosophy, or science. Moreover, Bacon is making bold promises as to the transformative nature of his project. Death, as part of the human condition, cannot be overcome, yet Bacon is careful to veil this truth in his presentation of Bensalem in *New Atlantis*. Is it not dangerous to gloss over such harsh realities? What is to stop the scientists of Solomon's House from abusing the power of science to control not only nature, but also their fellow men? These are questions that are not adequately answered within Bacon's presentation or within the works of his scientistic successors.

After exploring the crucial issues of scientism exhibited in Bacon's project, I briefly examine a momentous occasion in the history of modern science. While Bacon had called for a transformation of society based on science, he provided no concrete "results" of the new science's efficacy. It was not until Isaac Newton's work that science would gain the prestige necessary to transcend the bounds of intellectual circles into popular acceptance. Newton's work revolutionized physics, but had effects far beyond science itself. His work helped usher in a new worldview and led to unprecedented optimism within other intellectual fields, including the social sciences. It was assumed that Newton's *method* could be applied to other fields, with similar success. As an example of the attempted application of Newton's method to other fields, I briefly explore the project of the French *Encyclopedists*. Following in the footsteps of Bacon, and capitalizing on the success of Newtonian science, they sought to form a comprehensive document of human knowledge. Like Bacon, they rejected ancient and medieval philosophy and saw their own age as the true beginning of human knowledge. It was "to contain nothing less than the basic facts and the basic principles of all knowledge."[31] Reflecting the Enlightenment attitude, the Encyclopedists rejected traditional metaphysics and instead sought to base politics and ethics on the model of the natural sciences. Their program was based on a recasting of the educational system, something Bacon had also called for, and by the end of the eighteenth century their influence had been felt throughout Europe, and particularly in France.

Riding the wave of momentum created by the Enlightenment, their fellow countryman Saint-Simon sought to enact social reforms on scientific principles. Drawing on the prestige of Newton, Saint-Simon proposed the "Council of Newton" as the governing body that would enact his desired reforms. The scientists were to organize themselves along the lines of the Catholic clergy and direct the education of the masses. Furthermore, they would "elect a scientific pope, employ excommunication for crimes against the ideology, and institute a Newtonian form of baptism."[32] Thus, we find in Saint-Simon the desire for a complete transformation of society, not just the educational structure.

In spite of Saint-Simon's influence, he was never able to articulate a systematic vision of society based on "positive science." Instead, the task was carried out by his most prominent student: August Comte. Thus, I turn to a close examination of Comte's work in chapter four and demonstrate the scientistic character of his work. Comte sought to reorder every major institution from the family to business to the Church and ultimately to the regime itself.[33] He saw himself as the "Aristotle of Positivism" and expected his work to have the same type of influence that Aristotle's had; especially in Europe where the Aristotelians had been so influential in the Middle Ages.[34] Yet the disintegration of Christianity's influence and the rise of immanent materialism left more than just an intellectual void; the *spiritual* void created by the new worldview was immense. Thus, in addition to being the "Aristotle of Positivism" and establishing a new way to study man and understand society, Comte also had to mirror St. Paul and establish a new spiritual order. He did so by founding the Religion of Humanity as a replacement for Christianity. This study focuses on both major aspects of his work.

I begin with an examination of Comte's *Positive Philosophy,* as it is the foundation on which his later political and religious writings are based. Comte, like Bacon and the Encyclopedists, rejects ancient and medieval philosophy. His own work represents the third and final phase of human knowledge, and the validity of his work is based both on reason and on the dictates of history itself. The Law of the Three States is not meant to describe how things *should* progress, but rather how they *must* progress. The movement of history dictates that positivism will triumph, and Comte is merely the instrument that will bring it about.[35] Comte's main achievement in the first phase of his work (represented by his *Positive Philosophy*) is the establishment of "social physics." He seeks to do for social science what Newton's work did for natural science. Through the discovery of laws of human action, Comte seeks to provide solutions to the perennial problems of order that perplexed his predecessors and plagued societies throughout history. In addition to the establishment of social physics, Comte aims to unify all of the sciences in an attempt to further his social aims. Keeping with the utilitarian emphasis of his predecessors, he argues that *all* science should be directed toward the improvement of the human condition. Thus, he calls for "one more great specialty, consisting in the study of general scientific traits. These scientists will determine the exact character of each science and reduce its chief principles to the smallest number of laws or principles within the confines of the positive method."[36] The other scientists, before receiving specialized training, should first receive training in the general principles of positive knowledge. If these two conditions were fulfilled, "the modern organization of the scientific world would then be accomplished, and would be susceptible of indefinite development, while always preserving the same character."[37]

In spite of Comte's confidence in the development of positive philosophy and its ultimate triumph (decreed by history itself), he realizes that an intellectual revolution is insufficient to the task at hand. People need something to believe in, and reason, no matter how sound, is not sufficient to convince the masses. In line with his positive philosophy, Comte offers a "demonstrable" religion that is designed to supplant the "revealed" religion of Christianity. After serving as the "Aristotle of Positivism," Comte takes on the mantle of St. Paul in the founding of his new religion. To explore this facet of Comte's work, I focus on the *Catechism of the Positive Religion*. As evident in the title, Comte found Catholic organizational principles appealing, although this admiration did not extend to Catholic doctrine. As Aldous Huxley famously noted, Comte's positive religion can be characterized as "Catholicism minus Christianity."[38] The universality desired by the Catholic Church could truly be achieved through a demonstrable, scientific religion.[39] Religion had historically served as a source of order and Comte realized its utility in that regard. However, *progress* could not occur as long as that religion was stuck in the outdated modes of metaphysical and theological philosophizing. Thus, Comte set out to base his doctrine on positive principles that could presumably be accepted by everyone since they would be demonstrable. Faith in an unseen God would no longer be required; the only faith required would be in the science upon which Comte based his system. Yet, that faith was not based on hope, but was "guaranteed" by history itself since it was based on positive science.

Comte replaces the Christian God with a new deity, Humanity, which he defines as:

> The whole of human beings, past, present, and future.
> The word *whole* points out clearly that you must not

take in all men, but those only who are really assimilable, in virtue of a real co-operation towards the common existence. Though all are necessarily born children of Humanity, all do not become her servants, and many remain in the parasitic state which was only excusable during their education.[40]

The true servants of Humanity are rewarded through commemoration and veneration after their deaths, while the others are left to toil in eternal obscurity. Consistent with Baconian and Enlightenment thought, Comte's religion does not place primacy on the cultivation of the soul, but rather on physical health. This is evident in the incorporation of medicine into the positive priesthood (a strikingly similar position to the scientists of Solomon's House in Bacon's *New Atlantis*). Anyone that resists Comte's project can be characterized as a detriment to Humanity, and thus serious limits are placed on religious and political liberty within his system.

After the examination of Comte, I explore the scientistic foundations of one of modernity's most influential political thinkers: Karl Marx. While countless studies have been devoted to Marx, I focus on the foundational aspects of his work, and especially on his (anti) philosophical anthropology. My goal is to show continuity with his scientistic predecessors and to note the importance of the prestige of science to Marx's work. While Marx decries the social philosophies of Comte and Saint-Simon because of their "utopianism," Marx's own project rests on a dogmatic faith in history and science. Following in the anti-metaphysical spirit of Bacon, Comte, and the *Encyclopedists*, Marx focuses on *material* conditions and on the biological survival of man. Like Comte, he realizes the human need for spiritual fulfillment. While Comte turns to the religion of humanity, Marx points to communism. Marx realizes the skepticism that might follow from a critical examination of his system and thus explicitly prohibits questions about its premises.

One of the key premises that I focus on is Marx's theory of human nature: a theory that receives its justification by an appeal to history and science. According to Marx, the rise of natural science has allowed for man's true nature to be revealed. It has "invaded and transformed human life all the more practically through the medium of industry, and has prepared human emancipation, however directly and much it had to consummate dehumanization. Industry is the actual, historical relation of nature, and therefore of natural science, to man."[41] It allows us to understand "the human essence of nature or the natural essence of man."[42] Therefore, natural science will become "the basis of human science" just as it has become "the basis of actual human life, albeit in estranged form."[43] Man's "real nature" is revealed by human history and "hence nature as it comes to be through industry is true anthropological nature."[44] Natural science will in time subsume under itself the science of man, just as the science of man will subsume under itself natural science: there will be one science."[45] Man becomes the immediate object of natural science just as nature becomes the immediate object of the science of man.[46] Marx thus blurs the distinction between man and nature, and reduces the knowledge of both to sensory perceptions and history. Marx's prohibition of questioning follows directly from his understanding of man and nature. Man's experiential knowledge points to the fact that he is not a self-created being, yet that is precisely the premise that Marx needs to uphold for his system to work. Thus, his response to questions of human origins, of who begot the first man, is to simply reject the question. As Marx explains: "for the socialist man the entire so-called history of the world is nothing but the begetting of man through human labor, nothing but the coming-to-be of nature for man, he has the visible, irrefutable proof of his birth through himself, of his process of coming-to-be."[47]

The other crucial premise to Marx's system, which also bears scientistic prejudices, is that history has decreed the coming success of communism. Marx asks his followers to place faith in the eventual victory of communism. This is crucial because the transformation that Marx calls for will not happen until *after* a bloody, violent revolution. Marx's critique of social theorists such as Saint-Simon and Comte had nothing to do with the ends they sought, but the *means* with which they sought to accomplish them. He thought it was utopian and unrealistic to expect a radical transformation of society through philosophy or education alone. Therefore, unlike those utopian thinkers, "the Communists disdain to conceal their views and aims. They openly declare that their ends can be obtained *only by the forcible overthrow of all existing social conditions.*"[48] While the likes of Comte and the *Encyclopedists* had downgraded the value of the *bios theoretikos*, it is completely obliterated within Marx's system. Violent action is the means through which peace will ultimately be won and Marx abhors any "abstraction" that prevents that action from occurring. Moreover, like Bacon and Comte, Marx's vision is radically immanent. The paradise of the final state of communism will occur within history; thus, politics gains an absolute character. The important point for my analysis has little to do with fate of Marxism *per se*, but instead on the underlying assumptions and motivations behind its implementation.

The final demonstrative section of the study consists of a brief examination of Charles Darwin's evolutionary theory. Like Newton's *Principia*, Darwin's *Origin of the Species* helped shape a new worldview. Because of this fundamental shift, Darwin's theory has been influential in an array of subjects including biology, genetics, religion, criminal justice, psychology, philosophy, economics, and politics. However, like the analysis of Marx, I am concerned primarily with the underlying assumptions in Darwin's work. I seek to demonstrate how the reductionism inherent in Darwin's theory can lead to deleterious political consequences and also show how it is both the product of, and torchbearer for, scientism.

Darwin, like Newton, presents a materialistic worldview. And like Marx, he holds that human rationality, morality, and society are the "products of purely materialistic processes." Institutions such as religion and marriage are direct results of the evolutionary process.[49] The primary difference between Darwin's materialistic account and that of Newton is the lack of an appeal to design in nature. While the revelatory aspects of religion had certainly taken a hit with the rise of Enlightenment thought, natural theology had remained as an ally and perhaps last line of defense for Christianity. Under the natural theological view, the intricate calculations by the likes of Newton could be taken as a sign of perfect design. On Darwin's own account, his theory undermines such a view of the world: "the old argument from design in nature, as given by Paley, which formerly seemed to me so conclusive, fails, now that the law of natural selection has been discovered."[50] Darwin's discovery of natural selection leads him to abandon his own sense of wonder and awe at nature's workings: "I well remember my conviction that there is more in man that the mere breath of his body. But now the grandest scenes would not cause any such convictions and feelings to rise in my mind."[51]

In addition to the new view of nature, Darwin also presents a new view of man. Going against the traditional classical and Christian views of man that posited a special place for man based on his *differentia specifica*, Darwin seeks to minimize such differences between man and other beings. In the *Descent of Man*, Darwin demonstrates that while there are significant differences in degree, "the mental powers of the higher animals . . . are the same in kind with those of man."[52] Love, courage, revenge, suspicion, jealousy, pride, magnanimity, imitation, imagination, reason, and even a sense of humor can all be found in other animals.[53] Language can also be explained away through an appeal to materialistic processes: "language owes its origin to the imitation and modification, aided by signs and gestures, of various natural sounds, the voices of other animals, and man's own instinctive cries." Such a process may have originated with an "unusually wise ape-like animal" imitating the "growl of a beast of prey, as to indicate to his fellow monkeys the nature of the expected danger. . . . [T]his would have been a first step in the formation of language."[54] Darwin continues by dispelling the notion that consciousness and abstract thought are distinctly human and finally ends with an attack on the claim that "humans were distinguished from the lower animals by a universal belief in an all powerful creator."[55] Darwin points out the fact that "numerous races have existed and still exist who have no idea of one or more gods and who have no words in their languages to express such an idea."[56] Moreover, he rejects the need for a Creator and posits religion to be a mere stage in the evolution of the human mind.[57]

I argue that Darwin's reductionist view of man and nature can lead to disastrous political consequences. As a concrete example, I examine the eugenics movement and Nazi Race Theory; neither of which would have been possible without the new view of man proffered by Darwin. Francis Galton, Darwin's cousin, coined the term *eugenics* in 1883 and the basic assumption was that the "only solution to social problems was to discourage reproduction by those with undesirable traits, while encouraging reproduction by society's worthier elements."[58] While "positive" eugenics was attempted initially (i.e., encouraging those with desirable traits to reproduce), it soon became apparent that it would not be enough to curtail the preponderance of those who were passing on less than desirable traits. The only solution then would be through "negative" eugenics, whereby those with undesirable traits would be forcibly prevented from procreating. By the early twentieth century, the eugenics movement was flourishing within the United States and gained acceptance amongst prominent educators and politicians. Thirty states adopted sterilization laws by 1940 and restrictions were placed on marriage and immigration as well. The power of the movement even made its way to the Supreme Court, where Justice Oliver Wendell Holmes (an avowed Darwinist) fully embraced eugenics. In the 1927 case of *Buck v. Bell*, Holmes wrote the majority opinion and upheld Virginia's compulsory sterilization statute. The chilling decision allowed for the promulgation of similar laws throughout the country, and the sterilization rate increased dramatically as a result.[59]

The eugenics movement within the United States lost momentum largely because of World War II. The Nazis had also embraced an active eugenics program, but the horrific results led many to abandon eugenics; or at least distance themselves from its full implications. The creation of a super race was an overt goal of Nazi policy based on the new view of man that had emerged largely through the work of Darwin. To understand how this view emerged, I turn to Voegelin's analysis of the race idea. Voegelin contends that the race theory of the Nazis is "an image of destruction" and represents the decay of race theory in general.[60] In the introduction to *The History of the Race Idea*, Voegelin laments, "The knowledge of man is out of joint. Current race theory is characterized by uncertainty about what is essential and a decline in the technical ability to grasp it cognitively."[61] Such a deficiency is characterized by the change in the view of man; or in Voegelin's terms, the shift from a Christian primal image to a post-Christian image.[62] The shift is characterized by a diminishment of man's spiritual capacity, with an increased emphasis on biological factors. Moreover, the source of life itself is placed within the organism, diminishing the need for reliance on outside forces (i.e., God). The new image of man is radically immanent and material and biological factors gain prominence. Thus, it is but a small step, once these ideas are popularized by Darwin, towards an active eugenics program and ultimately to the destructive Nazi race programs.

Finally, after tracing the development of scientism from Bacon to present day, I offer a solution to the problem. The argument throughout the study rests on the assumption that a proper understanding of human nature is crucial to politics. Man is the proper subject of politics and any theory of politics that rests on a faulty understanding of man is doomed to failure. The problem of scientism is ultimately one of misplaced hope and misguided reason. Natural science provides man with great power, but it does not provide guidance as to how that power should be used. Only a science of human affairs can provide such answers. Yet, the very science that is supposed to answer those questions, political science, has not escaped the debilitating effects of scientism. Social science was borne out of the very environment that scientism helped shape. The ability to critique and form political judgments gave way to the desire for certainty and mathematical precision. The methods that were successfully used in the natural sciences were adopted by the social sciences, but with the mistake of not realizing, or acknowledging, that those methods were designed for different, and less complex, subject matter. As a corollary to the importation of the methodology of the natural sciences, philosophical questions were either discouraged or banned outright since they could not be answered with those methods. Modern political science is thus unable to address the crisis of scientism, not only because it is a product of it, but also because it fails to provide the diagnostic tools necessary to even recognize the problem.

My solution then points towards a new science of politics, one rooted in the philosophical anthropology provided by the likes of Plato and Aristotle, and kept largely intact by Christianity.[63] Man is the epitome of being since he participates in all levels of reality and a science of man must take that into account. While his corporeal nature is undoubtedly important, it must not be mistaken for the whole of his existence. *Apperceptive* experience must be reincorporated into the science of politics and the *bios theoretikos*, or contemplative life, must be taken seriously. Political science must be studied from the perspective of the statesman and citizen: not from an imaginary Archimedean point of neutral objectivity. This requires a change both in the spirit and language of political science. Political science should seek to provide guidance in concrete political matters. And if it is to successfully counsel the statesman and citizen, it must speak in terms they can understand. The desire to be *scientific* is misdirected if understood in the sense of a particular method and its terminology. Politics necessitates flexibility: both in thought and action and a proper science of politics must be reflective of that fact. Now that I have outlined the general aims of the study, I will turn to the specifics and explore the problem of scientism in depth.

Chapter 2
Francis Bacon and the New Science

"From these and all long Errors of the way,
In which our wandring Praedecessors went . . .
Bacon, like Moses, led us forth at last,
The barren Wilderness he past,
Did on the very Border stand
Of the Blest promis'd Land,
And from the Mountain Top of his Exalted Wit,
Saw it himself, and shewed us it."[1] –Abraham Cowley

Francis Bacon (1561–1626) is undoubtedly one of the great minds of modern thought. Thomas Jefferson included Bacon as part of his "trinity of the three greatest men that have ever lived, without exceptions."[2] Jean Jacques Rousseau referred to him as "possibly the greatest philosopher."[3] And John Dewey referred to him as the "real founder of modern thought."[4] Bacon was *the* partisan for the advancement of science within the early modern period. Although he cannot be credited with a particular scientific achievement,[5] Bacon helped establish the foundation on which the new science was to be built. He was the first to outline the scientific method, with an emphasis on experimentation and observation.[6] And perhaps more importantly, he argued for the adaptation of that method. The benefit of science was not readily apparent to his contemporaries, and it was no small feat to convince them of its utility.

While Bacon can arguably be credited as the founder of the new science and deserves praise for that feat, he also must be held accountable for the deleterious effects of his project. Bacon's dismissal of Plato and Aristotle, along with his seemingly unbounded optimism in regards to the transformative nature of his project, led him to overlook important aspects of political reality. While he was right to criticize the lack of progress within natural philosophy, he too readily dismissed the political and ethical dimensions of ancient and medieval thought.[7] Bacon substitutes his new science in the place of the natural philosophy of his predecessors, but fails to adequately account for politics. This is not to say that he ignores the subject altogether. Instead, he subsumes it under the umbrella of his new science:

> It may also be doubted (rather than objected) whether we are speaking of perfecting only Natural Philosophy by our method or also the other sciences, Logic, Ethics and Politics. We certainly mean all that we have said to apply to all of them, and just as common logic, which governs things by means of the syllogism, is applicable not only to the natural sciences but to all the sciences, so also our science, which proceeds by *induction*, covers all.[8]

In other words, his *method* can be applied to every facet of knowledge. The problem with this is at least twofold: man cannot be subjected to experimentation and his nature is not exhausted by sensory perceptions.[9] The second problem is related to the first in that some of the core questions of man's existence are off limits. The experimental method is designed to explain phenomenal relations and *how* things work. It does not and cannot answer questions of first causes or of "substance."[10] Furthermore, it does not provide guidance as to proper action, or ethics. Bacon explicitly denounces metaphysics and derides moral and ethical philosophy since it deals with the "proud knowledge of good and evil."[11]

A final troubling aspect of his thought is his utopianism. As evident in *New Atlantis*, Bacon essentially foresees no limit to man's ability to control his own fate through the domination of nature. He presents us with a technological society that seemingly knows neither death nor disorder—in other words, heaven on earth. The only precondition for this earthly salvation lies in the adoption of

Bacon's new science. Thus, man no longer needs to rely on Providence for assistance or pin his hopes on an otherworldly existence. Instead, he can manipulate nature to provide a seemingly endless array of earthly goods.

Bacon's dogmatic emphasis on method (and the postulate that it can be utilized in all areas of knowledge), along with his reductionist account of man and his utopianism, leads me to claim that he was not only the founder of the new science (as he is often credited, and rightly so), but also that he serves as the founder of *scientism*.[12] To support this claim, I will examine two of Bacon's most important works: *New Atlantis* and *Novum Organum*. *Novum Organum* provides insight into Bacon's project as a whole, as it gives us both a critique of the past and a plan for the present and future. *New Atlantis* gives us a vision of the "final cause" of Bacon's project. It shows Bacon's regime *in action* and is invaluable for understanding his politics.

Before delving into the specifics, it is important to note Bacon's place within the history of ideas. As one of the founders of modernity, Bacon's project is presented as a corrective to the politics of his ancient and scholastic predecessors. He likens himself to Columbus and his project serves as the vessel that will usher in a new era of peace and prosperity. Machiavelli had also compared himself to Columbus, yet the two offered strikingly different solutions to the political crises of their day. Machiavelli's solution depended on a strong prince, at least initially, who would take bold action in defending his people and expanding his empire. Military conquest was at the forefront and war was inevitable. On the other hand, Bacon's ideal society is one of peace and prosperity and the movement towards such a society is done through educational reform, not through warfare. Furthermore, the society's success is not dependent on the virtue of the prince, but rather on a group of technically trained scientists.

Fortune, or chance, is what ultimately separates the politics of Bacon and Machiavelli. Machiavelli constantly laments the inability of man to conquer nature. The virtue of the prince can certainly mitigate the deleterious effects of bad fortune, but even the most virtuous man cannot hope to completely conquer it. Thus, Machiavelli's politics is "limited" by nature. He never presents politics as anything other than a constant struggle for power and order. Even if the rare prince comes along and restores order, that order will only be temporary; as republics will eventually become corrupt and require another prince to step in. This cyclical view of history keeps Machiavelli in the realm of "realism" as he never proposes a society that cannot exist in practice. On the other hand, Bacon's proposed regime is "transhistorical" and is ultimately immune to the vicissitudes that Machiavelli lamented.[13] Baconian science serves as the remedy to the ills that have plagued man throughout history. "Whoever has insight into the nature of man may shape fortune almost at will . . . and is born for empire."[14] And by "vexing" nature, man learns to control it. Bensalem is an earthly paradise that offers the prolongation of life (if not eternal life) in addition to prosperity. Thus, Bacon's thought can be characterized as utopian.[15] It is now important to turn to the specifics.

Bacon's Method

Throughout his writings, Bacon employs a variety of methods to convey his thoughts. *Novum Organum* is presented through aphorisms. Bacon's use of aphorisms points to the fact that his project is a "work in progress" and is "susceptible of improvement and refinement."[16] As Bacon later explains in his *Advancement of Learning*:

> Delivery by aphorisms . . . tries the writer, whether he be light and superficial in his knowledge or solid. For aphorisms, not to be ridiculous, must be made out of a pith and heart of sciences. . . . A man will not be equal to writing in aphorisms nor indeed will he think of doing so, unless he feels he is amply and solidly furnished for the work.[17]

Bacon clearly thinks he is up to the task, but he also realizes the need for others to contribute. Although he may be writing in relative isolation, the success of his project demands a massive collaboration and his choice of aphorisms points to this realization. On the whole, *Novum Organum,* as part of the *Great Instauration,* is an expression of Bacon's explicit intentions. This makes sense given the collaborative nature of the project. Bacon notes, "If we have too readily believed anything, if we have fallen asleep or not paid enough attention, or given up on the way and stopped the inquiry too soon, we still present things plainly and clearly. Hence our mistakes may be noted and removed before they infect the body of science too deeply; and anyone else may easily and readily take over our labours."[18]

While Bacon explicitly expresses a desire for clarity in regards to the scientific method, he takes an altogether different approach when discussing politics. Bacon refers to politics as a "secret and retired" subject.[19] His *Essays* are marked with contradictions and cryptic references and clearly are not intended for "mass consumption." And arguably his most political work, *New Atlantis,* is presented via myth and is formally incomplete. Yet, Bacon's teachings on politics are indispensable to his project as a whole. A successful implementation of his science depends on the general acceptance of society as well as, and perhaps more importantly, the regime. Bacon clearly understands the importance of this dual acceptance and carefully caters his various works to this purpose. Thus, to comprehend Bacon's project as a whole, one must be able to penetrate the meaning of his political writings.[20] While not easy to do, Bacon provides the reader with a lighted pathway, albeit with a dimly lit candle rather than a torch.

Just as he did with his choice of aphorisms in *Novum Organum,* Bacon provides reasons for choosing to utilize a myth in *New Atlantis.* Bacon's own understanding of the role of myths evolved over the course of his writings. In the first version of the *Advancement of Learning,* he seems to dismiss interpretations of myths as coming solely from the reader noting, "I do rather think that the fable was first, and the exposition devised, than that the moral was first, and thereupon the fable was formed."[21] Four years later, in *The Wisdom of the Ancients,* Bacon backs off his initial position and acknowledges the possibility that the authors veiled certain truths. "I do certainly for my own part (I freely and candidly confess) incline to this opinion; —that beneath no small number of these fables of the ancient poets there lay from the very beginning a mystery and an allegory."[22]

Bacon's decision to utilize myth therefore deserves the reader's full attention. Considering his admission that myths can veil as well as instruct, one must ask what needed to be veiled and why this was so. One obvious answer is that he wanted to protect himself from the "old" guard. Bacon's call for sweeping changes in the way science is practiced may have been revolutionary, but it alone, provided no apparent threat to the existing authorities. A political science that calls for a new organizational structure to society is a different story altogether. For science to be truly advanced, society had to be conditioned to accept it. And this required a shift in politics. Bacon was well aware of this and thus took great care in presenting his thoughts on the "secret and retired" subject. A second, and related, reason is that Bacon had to be careful not to attack the Church.[23] While Bensalem is a place that seemingly embraces Christianity, there are several issues that bring into question the importance of Christianity to Bacon's scientific utopia.[24] If Bacon's vision had any chance of being successfully implemented, he could not afford to alienate the Church or the Christian population as a whole. A third reason gets back to the point of "unfinished" business. The myth itself is formally incomplete just as Bacon's project is seemingly left unfinished. While the vision of the end is there, the means have yet to be fully developed and will require successors to do so. Now that I have acknowledged the methodological approaches of Bacon and briefly examined the reasoning behind their implementation, it is time to turn to the works themselves.

Novum Organum

Novum Organum stands as Bacon's most important philosophical work. It comprises the second part of Bacon's six-part *Great*

Instauration. As such, it is designed to "equip the human understanding to set out on the ocean."[25] As Bacon notes:

> We plan therefore, for our second part, an account of a better and more perfect use of reason in the investigation of things and of the true aids of the intellect, so that (despite our humanity and subjection to death), the understanding may be raised and enlarged to overcome the difficult and dark things of nature. . . . Different results follow from our different design. . . . [W]e conquer nature by work.[26]

From the onset, Bacon makes it clear that his project is novel. He rejects the traditional modes of investigation, namely Aristotle's logic, for it proves to be "quite divorced from practice and completely irrelevant to the active part of the sciences."[27] According to Bacon, the syllogism fails to get at first principles and cannot provide the insight into nature that is required for true progress in the sciences. And "there has been no one who has spent an adequate amount of time on things themselves and on experience."[28] The solution lies in a new method, which will provide the proper tools to aid human understanding. "Before one can sail to the more remote and secret places of nature, it is absolutely essential to introduce a better and more perfect use and application of the mind and understanding."[29]

To get to the "secret places of nature," Bacon offers his experimental method. The experiment provides an "assistant to the senses" and therefore:

> We do not rely very much on the immediate and proper perception of the senses, but we bring the matter to the point that the senses judge only of the experiment, the experiment judges of the thing. Hence we believe that we have made the senses (from which, if we prefer not to be insane we must derive everything in natural things) sacred high priests of nature and skilled interpreters of its

oracles; while others merely seem to respect and honour the senses, we do so in actual fact.[30]

Bacon's method is designed to overcome the problems that have plagued the advancement of natural philosophy up until his time. He claims not to be "dethroning the prevailing philosophy" or disrespecting the achievements of the ancients, but merely guiding us to the correct path. This somewhat guarded language in the preface gives way to a sharper tongue as Bacon begins Book I.

Book I focuses on the inadequacy of the current state of knowledge and the need for a renewal of learning. Bacon asserts, "no great progress can be made in the doctrines and thinking of the sciences, nor can they be applied to a wide range of works, by the methods commonly in use."[31] In fact, he points to only three periods in history where actual progress was made in natural philosophy: the Greeks, Romans, and present day Western Europeans.[32] And those periods only saw limited progress. There are numerous reasons for the lack of progress, but Bacon specifically mentions four "illusions of the mind" that are particularly troublesome.[33] These are the idols of the tribe, idols of the cave, idols of the marketplace, and idols of the theatre. The idols of the tribe "are founded in human nature itself."[34] This includes the influence of the emotions and limitations of the senses. The idols of the cave "have their origin in the individual nature of each man's mind and body; and also in his education, way of life and chance events."[35] This includes the inclination of individuals to admire tradition or to embrace novelty. The idols of the marketplace are the "biggest nuisance of all, because they have stolen into the understanding from the covenant on words and names."[36] In other words, language is problematic. Words need to be carefully defined and must refer to observable objects.[37] Finally, the idols of the theatre are "openly introduced and accepted on the basis of fairytale theories and mistaken rules of proof."[38] Bacon includes religious and philosophical sects in this group.[39] All of these idols must be "rejected and renounced and the mind totally liberated and cleansed of them, so that there will be only one entrance into the kingdom of man, which is based upon the sciences, as there is into the kingdom of heaven."[40]

Aside from the idols of the mind, "the greatest obstacle to the progress of the sciences and to opening up new tasks and provinces within them lies in men's lack of hope and in the assumption that it is impossible."[41] Indeed Bacon's reliance on hope is critical to the flourishing of his project as a whole. We must have a reasonable expectation that the new path recommended by Bacon will lead to tangible benefits. Therefore, Bacon states: "we should reveal and publish our conjectures, which make it reasonable to have hope, just as Columbus did, before his wonderful voyage across the Atlantic Sea, when he gave reasons why he was confident that new lands and continents, beyond those previously known, could be found; reasons which were at first rejected but were afterwards proven by experience, and have been the causes and beginnings of great things."[42]

The primary reason for hope is that the lack of progress in the past has been due to human error. Men simply focused on the wrong things and did not devote their time properly to the study of natural philosophy. Bacon notes, "every error that has been an obstacle in the past is an argument of hope for the future."[43] Furthermore, "Natural philosophy is not yet found in a pure state, but contaminated and corrupted: in the school of Aristotle by logic, in the school of Plato by natural theology. . . . Better things are to be hoped from natural philosophy pure and unadulterated."[44] Bacon is the one who will accomplish this feat and he fully expects posterity to reward him for it: "if someone of mature age, with faculties unimpaired and mind cleansed of prejudice, applies himself afresh to experience and particulars, better is to be hoped of him. And in this task we promise ourselves the fortune of Alexander the Great; and let no one accuse us of vanity before he sees the result of the thing, which aims to uproot all vanity."[45]

Bacon again emphasizes the feeble state of natural philosophy, but he sees an opportunity to forever change its course through his experimental method. Unlike the past where nature was merely observed, Bacon's method requires active participation and manipulation of nature: "Just as in politics each man's character and the hidden set of his mind and passions is better brought out when he is in a troubled state than at other times, in the same way also the secrets of nature reveal themselves better through harassments applied by the arts than when they go their own way."[46] Since this has not been tried, "it is very much to be expected that many exceedingly useful things are still hidden in the bosom of nature . . . which have however not yet been discovered, but without a doubt will appear sometime."[47] Since "men have not spent much time on experience" we have not yet reaped the rewards that surely await us. Bacon goes so far as to say, "if there were anyone present among us who would answer interrogators about the facts of nature, *it would only take a few years to discover all causes and all sciences.*"[48] Bacon does not promise to deliver these results himself since he is "the busiest man" of his age in political affairs and "not in very good health."[49] In spite of his limited time and opportunity, Bacon has managed to outline the method that can bring about such discoveries. Therefore, he concludes that we should have "an abundance of hope."[50]

The theme of hope is one that Bacon sticks with throughout his works and it finds its greatest expression in the *New Atlantis*. Before turning to that work, it is helpful to note a few passages in the *Novum Organum* that seem to be at odds with what is presented in the *New Atlantis*. Although Bacon implores us to have hope, he chides those "glib, fanciful talkers" who promise more than they can deliver:

> Promising and advertising longer life, postponement of old age, relief from pain, healing of natural defects, temptations for the senses, enchantment and excitement of the passions, stimulation and enlightenment of the intellectual faculties, transmutation of substances, unlimited power and variety of movement, impressions and alterations of air, drawing and control of celestial influences, divination of future things, representation of distant things, revelation of hidden things and much more

of the same. The right verdict on these false benefactors is that in philosophical teaching there is just as much difference between their empty promises and the true arts as there is, in the narratives of history, between the achievements of Julius Caesar or Alexander the Great and the deeds of Amadis of Gaul or Arthur of Britain.[51]

Bacon claims that these "impostors" have caused prejudice against novel claims, thus making his task even more difficult. What is not clear is whether Bacon finds the above promises to be impossible outright or whether it is simply a function of them not following the correct method. The answer to that question is critical, especially in light of what Bacon says in the *New Atlantis*.

A final passage worth noting can be found in the *Great Instauration*. After Bacon has outlined his six-part plan, he prays, "May God never allow us to publish a dream of our imagination as a model of the world."[52] Four years later, he wrote *New Atlantis*, his last major work and one of the great myths of modern times. Does this work violate the principle outlined in the *Great Instauration*? Has he indeed published a dream of his imagination as model of the world?[53] This gets back to the question of what is possible with the correct method and what is simply impossible. To answer this question, and to better understand his work as a whole, I must now turn to the *New Atlantis*.

New Atlantis

Of all Bacon's writings, the *New Atlantis* is arguably the most controversial. This is not surprising considering both the form of the work and its content. I have already discussed the general reasons behind the implementation of the myth, but now must delve into further details. The formal incompleteness of the work seemingly has a simple answer: Bacon did not have time to complete it before his death. This is essentially the argument set forth by Rawley, Bacon's secretary who oversaw its publication.[54] Yet some, like

Jerry Weinberger, argue that the work is not as incomplete as it appears. Weinberger notes:

> We must consider that the New Atlantis may indeed be the picture of the end of science as anticipated in the sixth part of the plan, and that it is a *complete* picture that appears incomplete because it presents a "secret and retired" teaching about politics that can be discerned only with difficulty.[55]

Moreover, the *New Atlantis* evokes Plato's story of Atlantis, told in the *Timaeus* and *Critias*.[56] Plato's story is also formally incomplete as it breaks off before Zeus's speech.[57] This points to an intentional reason for the myth's incompleteness as opposed to it merely being an accident of time and circumstance. And the fact that Bacon went through the trouble to have it published in Latin serves as more evidence of its importance to his project.

While the form is obviously important, the main concern must be with the content. More specifically, I will focus on the political aspects of the myth with a particular emphasis on its *scientistic* character. I will show how Bacon violates his own principle by publishing "a dream of our imagination as a model of the world."[58] His seemingly unbounded optimism in the power of science led him to overlook the most pressing issues of human existence. The sobriety found in his earlier writings, in which he constantly warns against forgetting our mortal condition, gave way to an intoxicating dream in his final great work. Bensalem is essentially an earthly paradise, which is insulated from the very things that have plagued every other society in history. The society itself serves as redemption for the human race, and Bacon's project is the vessel by which it can be accessed. Now, I will turn to the specifics in order to demonstrate exactly why it is indeed "a dream of our imagination" and more importantly, why it is a *dangerous* dream.

The basic plot of the *New Atlantis* sees a group of European sailors land on a mysterious island. They had been lost at sea with little hope of finding their way back and a number of them were sick. Their prayers for land were finally answered, but they were not immediately admitted onto the island. Only after they proclaimed themselves to be Christians and swore that they had not engaged in warfare for the previous forty days, were they allowed onto the island and placed in the Strangers' House. The opening scene is important for several reasons. The first is that Bacon employs a recurrent theme of navigation.[59] In addition to fitting in with the literature of his time, it serves as a symbol of how Bacon views his own project. Having likened himself to Columbus, Bacon is offering entrance into a new world. While hardships are not entirely absent from the journey to Bensalem, Bacon minimizes their importance. He does not describe deaths of crewmen or terrible storms. The winds had apparently been favorable for the first five months of the trip, but then shifted to where they could no longer sail in their desired direction. This rather peaceful view of the journey can be likened to the transition that will have to occur in society as Bacon's science is accepted. In other words, Bacon holds out hope for a rather peaceful transition even though such transitions throughout history have been riddled with violence.

It is important to note the conditions for coming onto land as well. They both have to do with religion. The sailors are not afraid to profess their Christianity (in light of seeing a cross) and the officials there to receive them are pleased to hear so. Moreover, the oath of "peace" for the previous forty days plays on Christian and Judaic symbolism and further confirms the nature of Bensalem as a peaceful society. The sailors are impressed with the generosity of their hosts and express a desire to stay there permanently after only a short time on the island. Yet the seemingly hospitable nature of the hosts may not be quite what it appears. For it is soon revealed that Bensalem possesses laws of secrecy that forbid guests to visit the island and then impart their knowledge of it upon returning to their homelands. The Governor of the House of Strangers insists that none have been held against their will, but for such laws to be successful, it seems almost necessary that it would have to be done in cases where visitors refused to stay. He claims those returning from the island would not be believed even if they did divulge information upon their return, but if this is truly the case, then there would be little reason to try to convince visitors to stay in the first place.[60]

Nevertheless, in this case, the sailors express a genuine willingness to stay; mainly as a result of their favorable impressions of the Strangers' House during the mandatory three-day stay:

> So we spent our three days joyfully, and without care, in expectation what would be done with us when they were expired. During which time, we had every hour joy of the amendment of our sick, who thought themselves cast into some divine pool of healing, they mended so kindly and so fast.[61]

Indeed, the emphasis on healing and prolongation of life is one of the primary aims of Solomon's House. While death is not mentioned anywhere in the story, it is important to note that there are fewer chambers for those who have healed than those who are sick. This implies that not everyone will be healed. In other words, it confirms our mortal condition, but Bacon conceals this fact and instead emphasizes the seemingly miraculous recovery of the sick sailors.[62]

It is through the governor's speech that we also discover something about the history of Bensalem. The first question asked by the sailors is how the island came to be converted to Christianity. The governor notes that twenty years after Christ's ascension, a great pillar of light appeared off the coast of Renfusa. One of the wise men from Solomon's House was present and after declaring the light to be miraculous, the wise man's boat was allowed to approach closer (all of the other boats had been stopped within sixty yards of it). As he got close, the pillar of light broke up, but a small chest was left behind. It contained all of the books of the Old and New Testaments, in addition to "some other books of the New Testament, which were not at that time written."[63] The most striking part of the account is that the wise man declared the pillar of light to be a miracle: reason judges revelation. But what could give him such authority to distinguish a miracle from a fraud or a merely natural occurrence? The probable answer is Baconian science since Bensalem was founded well before the coming of Christ.

The fact that Bensalem was established well before Christianity is interesting for several reasons. For one, it shows that Bensalem's success was not due to Christianity. Solamona, the lawgiver of Bensalem, is described as a "divine instrument" who had a "large heart, inscrutable for good, and was wholly bent to make his kingdom and people happy."[64] Considering his wish for continued prosperity, Solamona instituted the laws of secrecy to preserve the exemplary nature of his society. He realized there were "a thousand ways" to make it worse, but "scarce" any way to make it better.[65] In other words, Bensalem was good from the start. It did not need anything else to make it better, but had to guard against corrupting influences to insure that it stayed that way.

So why did Bensalem convert to Christianity? First, it should be noted that the claim to conversion is not the same as actual conversion. Christianity has been admitted into Bensalem, but it is not clear how much influence it actually has had in the society. Nor is it clear how many citizens are actually Christian. Moreover, there are symbols from Islam, Judaism, and ancient Egypt present, and one of the most prominent characters in the book, Joabin, is Jewish. Secondly, it is important to remember Bacon's audience. He had to win over the churches or at least assure them that his project was not a threat. Thus, making the island compatible with Christianity was of utmost importance. But Bacon's timeline makes it clear that the success of the society did not hinge on Christian revelation; rather it was derived from science.

The religious symbolism found throughout the myth has been the subject of several interesting studies. Howard White and Jerry Weinberger both point to the symbols as being tools of the regime.[66] They serve to promote cohesiveness and order within the society. Stephen McKnight dismisses that claim as cynical and argues that Bacon's religious symbols were the reflection of genuinely held beliefs.[67] McKnight notes, "Bacon's vision of reform or instauration is drawn from the Judeo-Christian scriptures, particularly the

Genesis account of the Creation and the fall; from apocalyptic expectation of renewal in the Old Testament; and from soteriological themes of the New Testament."[68] Moreover, Bacon's work is also influenced by "themes and imagery found in the *prisca theologia*, a highly elastic collection of Neoplatonism, Hermeticism, alchemy, magic, and Jewish esoteric traditions."[69] Far from being a rejection of Christianity or religion, McKnight argues that Bacon's project aims to uncover a "truer, deeper level understanding of the scriptures and of God's saving acts in history."[70] The interpretations, while divergent, both offer valuable insight into Bacon's project.[71] Bacon is not merely trying to create order, although that is indeed a primary concern of his. He is attempting to create an earthly paradise and the religious symbolism is appropriate to his task. Bacon genuinely believes that his project can serve as redemption for mankind. Thus, the appropriation of Christian symbolism serves the dual purpose of propping up the regime *and* of reflecting Bacon's beliefs about the transformative nature of his science.[72]

Throughout the work, we are mainly relegated to getting information about Bensalem through its official representatives. An exception is the Feast of the Family, a public ceremony that is given to any man "that shall live to see thirty persons descended of his body alive together, and all above three years old."[73] The cost is incurred by the state. Since there is no mention of any specific virtue of the man, or of any further moral requirements, the feast appears to celebrate "mere longevity and fecundity."[74] This fits in well with the primary aims of Solomon's House: prolongation of life and material comfort.

After observing the Feast, the narrator meets one of the most important characters in the myth, Joabin the merchant. Joabin is described as a "wise man, and learned, and of great policy, and excellently seen in the laws and customs of that nation."[75] After the narrator praises the Feast of the Family, he inquires into the nature of marriage within Bensalem. Joabin explains how the laws and customs of marriage are arranged and why they are superior to those of Europe. Joabin claims that Bensalem is "the virgin of the world" and its people are of a chaste mind.[76] Unlike in Europe, marriage is not merely a convenient arrangement designed to quell "unlawful concupiscence." Instead, it represents a "faithful nuptial union of man and wife."[77] Joabin explains that there is no polygamy and at least a month must pass between the first meeting and marriage. The consent of the parents must be given and if it is not, then the inheritance is greatly diminished. Furthermore, Joabin tells us of the rejection of the idea of letting the married couple see each other naked before the contract. Instead, a friend of the man and a friend of the woman are allowed to watch them bathe in "Adam and Eve pools." This is a more "civil" solution because it avoids the "scorn to give a refusal after so familiar knowledge."[78]

The rites of marriage tell us more about Bensalem than may be apparent in a cursory reading. It marks one of the only passages where Bacon explicitly criticizes a European institution.[79] He is also leveling a critique against Plato and More since both had suggested the idea that married couples could see each other naked before marriage. But why is this distinction important? We could imagine that complications could arise from the arrangement in Bensalem. What if the friends became attracted to the spouses? Would not there be incentive to misrepresent what was seen in the "Adam and Eve" pools? Bacon seemingly ignores the eroticism that is inherent in these situations. As White notes, "the friend would have to overcome the sense of shame, and see them as they were before the fall."[80] McKnight argues that Bacon's wish is to restore man to his pre-lapsarian condition, and the presence of the Adam and Eve pools could suggest that his science has indeed been able to do that for Bensalem.[81] What is apparent is that marriage in Bensalem is not primarily a private good. Free choice is minimized.[82] The requirement of parental consent, with a significant inheritance penalty if not followed, and the implementation of the Adam and Eve pools suggest significant regulation of marriage by the regime. Thus, one can conclude that marriage serves a public good and this is reinforced in the aforementioned Feast of the Family, which could not be celebrated without a strong marital institution.

The final scene of the myth is also one of the most important for it allows us to see exactly what Baconian science can achieve.

As previously noted, the utopian aim of Bacon's project is to conquer chance. And that is exactly what is done through the workings of the scientists of Solomon's House in Bensalem. While Bacon acknowledges the importance of experiments of light, the true glory lies in the experiments that bear fruit.[83] Inventors are therefore revered since their works can benefit all of mankind and not just through the span of their respective lives. Their inventions live on well after they have passed.[84] The most important function of Solomon's House is the preservation and prolongation of life and so a great emphasis is placed on medicine.[85] We know that the medicine serves its function well from the account of the sick sailors who were healed in the "divine pool," presumably by the "Water of Paradise" referenced by one of the fathers of Solomon's House.[86] We learn that the scientists of Solomon's House have made all sorts of discoveries and have mastered the ability to manipulate nature to man's needs. They have genetically engineered beasts and birds:

> By art likewise we make them greater or taller than their kind is, and contrariwise dwarf them and stay their growth; we make them more fruitful and bearing than their kind is, and contrariwise barren and not generative. We find means to make commixtures and copulations of divers kinds, which have produced many new kinds, and them not barren, as the general opinion is. . . . Neither do we this by chance, but we know beforehand of what matter and commixture, what kind of those creatures will arise.[87]

In addition to genetic engineering, they have discovered flight and perfected machines that can imitate the motions of living creatures.[88]

Aside from the astonishing advances of Solomon's House, Bacon provides insight into its organizational structure. It is essentially a technological bureaucracy. The Merchants of Light are responsible for travelling, under concealment, to other lands to collect valuable information. The Depredators collect all known experiments from books. The Mystery-men collect experiments of "all mechanical arts, and also of liberal sciences, and also of practices which are not brought into arts." The Pioneers try new experiments as they see fit. The Compilers catalogue the experiments of the Pioneers. The Benefactors look for the practical use in those experiments. After a meeting of "our whole number" the Lamps direct new experiments that delve deeper into nature than the previous experiments did. The Inoculators carry out these experiments and finally, the Interpreters of Nature, "raise the former experiments into greater observations, axioms, and aphorisms."[89]

After the reader is given a glimpse of the hierarchy of Solomon's House, Bacon outlines an additional function:

> We have consultations, which of the inventions and experiences which we have discovered shall be published, and which not; and take all an oath of secrecy for the concealing of those which we think fit to keep secret, though some of those we do reveal sometime to the state, and some not.[90]

This is clearly a significant passage for it admits of political necessity. Technology wields great power, and that power can be used for good or bad. Thus, it becomes crucial to distinguish between helpful and potentially harmful discoveries. The reasons for such concealment are fairly obvious, but Bacon does not tell the reader *how* the decision is made to allow or to conceal inventions. One possible answer would be through *phronesis*, or the practical wisdom of a statesman, but it is unclear that the scientists of Solomon's House would possess it. Bacon does not even list politics as one of the subjects that is studied in Solomon's House so one must wonder from where political wisdom is to be derived.

Finally, the father of Solomon's House explains the most impressive feat of the institution, control of vicissitude:

> And we do also declare natural divinations of diseases, plagues, swarms of hurtful creatures, scarcity, tempests,

earthquakes, great inundations, comets, temperature of the year, and divers other things; and we give counsel thereupon, what the people shall do for the prevention and remedy of them.[91]

The science of Bensalem has overcome chance and has insulated the society against natural disasters and divine revenge. With that final revelation, the father of Solomon's House gave his blessing to the narrator and gave him permission to publish it "for the good of other nations."[92] The final line further confirms the contention that Bacon viewed the work as important and sufficient for publication. It also reinforces my characterization of his work; Bacon clearly thought that his project would benefit mankind as a whole and not just his particular society. The universalism of Baconian science transcends cultural and religious divisions (as evident by the harmony in Bensalem) and provides an earthly paradise similar to the one promised by Christianity, only in the afterlife.

Conclusion

From the analyses of the *Novum Organum* and *New Atlantis,* several important themes have emerged. The first is the rejection of the ancients, especially the Greeks, and scholastics.[93] Bacon finds the current state of natural philosophy to be wholly inadequate and places the majority of the blame on Plato and Aristotle, along with their respective followers. Bacon sees himself as lighting a torch "in the dark days of philosophy."[94] In the *New Atlantis,* this is symbolized by the journey to Bensalem, as Bacon's philosophy is the vessel by which man's estate can be improved. Furthermore, it points to an improvement of the Platonic myth, whereby the *New Atlantis* does not have to suffer the fate of the original since it possesses Baconian science and can protect itself against bad fortune.

Another important theme is the novelty of Bacon's project. In the realm of natural philosophy, he deserves credit for developing and promoting the experimental method. As such, he is rightly recognized as one of the founders of modern science. In the realm of politics, Bacon's innovation lies in his reliance on *hope*, as opposed to fear.[95] Bacon realizes that his project must overcome numerous obstacles if it is to be successful, and thus he must be able to convince his audience that his science will lead to tangible benefits. In *Novum Organum,* he states that this can be accomplished in part by looking at the particulars found in his table of discoveries. These can be taken as "an interest payment for the time being until the capital can be had."[96] In *New Atlantis*, he goes a step further and shows us what will be accomplished once that capital has been acquired. Yet, the question arises as to how the transition will occur from Bacon's contemporary society to one that mirrors Bensalem. The symbolism in the story suggests the transition will be peaceful and relatively smooth. However, such a transition would be essentially unprecedented as major transformations in society are almost always accompanied by violence. Bacon assures us that "the danger of not trying and the danger of not succeeding are not equal since the former risks the loss of a great good, the latter of a little human effort."[97] Indeed, that maxim may be considered the key claim of utopian thinkers. Given what occurred in the twentieth century, can such a claim be accepted today? Horrible things can indeed occur in the attempt to transform society regardless of whether the end is ultimately achieved or not. Tens of millions of people lost their lives at the hands of their own governments during the twentieth century.[98] The Holocaust and Gulags were implemented as tools to bring about an ideal society. Marxism promised a world of peace, but the means to getting there was through violent revolution. Likewise, the Nazis used the aforementioned pseudo-scientific race theory to justify the slaughter of millions of Jews, all in the name of progress. This is not to suggest Bacon would have endorsed either event. Nevertheless, the utopian principle that he presents is dangerous and can lead to less than desirable consequences.

Bacon's failure to account for the potentially destructive effects of his project (or at least the attempt to implement it), points to a larger problem within his work. In spite of his own extensive experience in politics, where he rose to one of the most powerful political positions in all of England, Lord Chancellor, Bacon seems to minimize the importance of political science. As mentioned earlier, he does not even include it in the subjects studied in Solomon's House. And he derides classical political science through his critiques of Plato and Aristotle in *Novum Organum* and the *Advancement of Learning*. In *Novum Organum,* he seems to suggest that politics can be studied using the same method set forth for the natural sciences. Yet, as noted earlier, that assumption proves to be problematic since human beings cannot ethically be subjected to experimentation and because their experiences extend beyond mere sensory perceptions. Towards the end of Book I of *Novum Organum,* Bacon anticipates a critique of his vision: "if anyone objects that the sciences and arts have been perverted to evil and luxury and such like, the objection should convince no one. . . . Just let man recover the right over nature which belongs to him by God's gift, and give it scope; *right reason* and *sound religion* will govern its use."[99] While this may be true, it must be asked *who* will exhibit right reason and *how* this can be assured. Without a program for political education, it seems dubious that "right reason" will be exhibited. This is why Plato and Aristotle went to great lengths to emphasize the importance of political education. Even if it is assumed that a society like Bensalem can exist, one still must wonder what guides the decisions of the scientists from a political and ethical standpoint. What standard is used to decide whether a particular invention or discovery will be disclosed to the public, or even the state? And who decides how it can be used and by whom? These questions cannot be adequately addressed without some sort of political knowledge, but Bacon gives us little guidance on the issue.

The fact that science has progressed as far as it has makes a satisfactory political science even more necessary. Our control over nature has vested us with great power. And with that greater power must come greater responsibility. Yet, political science in its current state is unable to offer much guidance. This is because of the deleterious effects of scientism on the social sciences. Too much emphasis has been put on the *method* used to acquire knowledge and certain forms of knowledge, particularly ethical and political, are discounted completely unless they conform to the methodological expectations of positivistic science. This type of attitude has all but erased metaphysics from the curriculum of political education, and the decline of moral philosophy can be traced to the same source. Perhaps "sound religion" is the corrective to that decline, but how can we expect to have sound religion when theology has also been degraded? *New Atlantis* points to a civil religion, but Bacon leaves little direction as to how it is to be formed and what principles should guide it. It is clear that Bacon himself possesses a strong moral sense, as his project ultimately aims at the relief of man's estate and on charitable and beneficial inventions. Yet, he derides the moral philosophy of the classics and scholastics and fails to offer a suitable replacement. Science may be able to provide man with the power to control nature, but it does not tell him how to use that power. Only a science of man, freed from scientistic reductionism, can provide us with the wisdom that is necessary to make such decisions. Bacon was prescient in his realization of the magnificent power that could be derived from science. And advances in medicine and technology have indeed helped to relieve man's estate. Yet, if science is to be restored to its rightful place, political science must first return to its rightful place as the master science of human affairs.[100] Before outlining what such a science must encompass, I must first explore some of the major developments in politics and science from Bacon's time to the present. My main focus will be on how the developments in science have impacted politics, with a particular influence on the concurrent rise of scientism. The first major development took place with Bacon's great English successor, Sir Isaac Newton, and it is to his work that I will now turn.

Chapter 3
Let There Be Light: The Newtonian Age

"Nature and Nature's laws lay hid in night; God said, Let Newton be! And all was light."[1]

While Francis Bacon deserves credit as a founder of modern science, the true impact of the scientific revolution was not felt politically until Isaac Newton. As Eric Voegelin notes, "The advancement of the science for which Newton is the great, representative genius has profoundly affected the political and economic structure of the Western world."[2] Indeed, Voegelin cites the advancement of science after 1700 as the "most important single factor in changing the structure of power and wealth on the global scene."[3] That advancement can be tied directly to Newton's two major works, the *Principia* and the *Opticks*. Newton's magnum opus is inarguably the *Principia*, or Mathematical Principles of Natural Philosophy.[4] It revolutionized physics and helped lay the foundation for a broad and cooperative scientific research agenda. More importantly, it changed the way that the world, and man's place in it, was viewed. To use Thomas Kuhn's language, Newton's work ushered in a new paradigm.

Newton was undoubtedly a scientist *par excellence*, and his philosophical works were often tempered by a sense of humility. Part of the impact of Newton's work was the separation of physics and philosophy, yet Newton himself operated within the realm of both.[5] He was careful to distinguish between the two, but he realized the importance of asking philosophical questions, even though he knew they could not be answered with any level of certainty akin to those given in the physical sciences. Newton's success created a sense of hope and optimism within European intellectual circles. It was thought that his *method* could be applied to other areas of knowledge with equal success. For instance, David Hume explicitly attempted to introduce the experimental method of Newton into moral subjects.[6] While Bacon had roughly outlined a new, unifying method, Newton's method was the first to actually garner the widespread support of practicing scientists and laymen alike. We must ask *why* it was assumed that Newton's method could be applied to other fields of knowledge, as the likes of David Hume had assumed. The answer lies in the *Principia* and it is to that work that I now must briefly turn.

Principia

The *Principia* stands as one of the greatest scientific works in human history. Its influence on science is indisputable, but some have overlooked just how much it impacted Western thought as a whole. Newton's work extends beyond the intellectual into the spiritual and political. It revolutionized physics, but also brought a whole new worldview into focus. The *Principia* mathematized the physical world and outlined immutable laws of nature. Newton famously refused to offer hypotheses in the same way his contemporaries and predecessors had: "hypotheses non fingo."[7] In refusing to feign hypotheses, Newton meant to avoid making statements that could not be supported by independent empirical support: "For whatever is not deduced from the phenomena must be called a hypothesis, and hypotheses, whether metaphysical or physical, or based on occult qualities, or mechanical, have no place in experimental philosophy."[8] While Newton himself did offer hypotheses and found metaphysical questions to be of interest, he was careful to separate those from his scientific findings.[9] He was reacting against the likes of Descartes and Leibniz; both of whom he thought had let metaphysical suppositions interfere with their natural philosophy.[10]

While Newton continued to realize the importance of metaphysical questions himself, the effect of his work was to further minimize its importance.[11] As Voegelin notes, the *Principia* became *the* representative document for the materialization of the external world. "By materialization of the external world we mean the misapprehension that the structure of the external world as it is constituted in the system of mathematized physics is the ontologically real structure of the world."[12] Voegelin continues:

> The impact of this masterful systematization of mechanics on his contemporaries, coming at a time when the sources of an active faith were drying up, must have had a force that is difficult to reproduce imaginatively today. To a spiritually feeble and confused generation, this event transformed the universe into a huge machinery of dead matter, running its course by the inexorable laws of Newton's mechanics. The earth was an insignificant corner in this vast machinery, and the human self was a still more insignificant atom in this corner.[13]

Voegelin finishes his analysis by noting that the *Principia* "is at least as important as the cause of the great schism in Western thought as it is important in the advancement of science."[14]

What is important to note from Voegelin's analysis is the *spiritual* dimension to Newton's work.[15] The decline of Christianity and rejection of Scholasticism had led to a spiritual crisis. The void that had formerly been filled by God could now be filled by science. "A new world-filling reality, emerging from Galilean and Cartesian physics and systematized in Newtonian mechanics, is ready to substitute for God and his creation."[16] God becomes a mere hypothesis and as the "mechanism of matter extends infinitely. . . . God has been squeezed out of his world."[17]

It is instructive to recall some of the key features of scientism and demonstrate how Newton's work fits into that framework. The Royal Society, of which Newton became the great representative figure, was in large part modeled after Solomon's House.[18] Bacon would have been pleased to know that such an influential organization would be formed, whereby science could be advanced through a cooperative effort of scientists. However, one should be careful not to ascribe to Newton the same attitude that Bacon carried in regards to the transformative nature of natural science. Newton did not overtly prescribe anything like Bacon did in the *New Atlantis* nor did he intend for his work to displace Christianity or to serve as a new spiritual foundation for society.[19] Even though he may not have intended to do so, Newton's own work served to at least partially bring about the type of transformation that Bacon had yearned for, but failed to achieve through his own works. However, the explanation for Newton's success conveniently fits into Bacon's scheme. As mentioned previously, Bacon could not be credited with a single scientific achievement. On the other hand, Newton had countless achievements.[20] Newton gained such respect because of the *utility* of his work. Bacon's emphasis on useful knowledge had permeated Europe by the time of Newton. Thus, it came as no surprise that Newton would be revered for his tangible results and could bring about a transformation that Bacon had only dreamed of.

Newton's astonishing success led to reverence among many of his peers, as well as predictable jealousy from rivals, and rightly so. Yet, some failed to realize the limits of Newton's work, even though he was quick to point out his inability to answer certain questions.[21] The inability of Newtonian science to answer those questions did not lead to skepticism. Instead, the questions that could not be answered within Newton's framework were to be considered meaningless, or at least not worthy of serious attention. This assertion was the necessary result of elevation of Newton's *method*. Bacon's emphasis on method had indeed been influential and the success of Newton's method to answer questions that had thereunto been unanswerable, led to a dogmatic acceptance of that method. There were at least two consequences that followed: the belief that the method could successfully be applied to all realms of knowledge and whatever could not be examined with the method became irrelevant.

The assumption that Newton's method could be applied to all

realms of knowledge was mainly a result of the mechanistic world-view presented in the *Principia*. The view of the world as a machine, with inexorable laws, led to the widespread acceptance of quantitative methods. While Newton may not have been able to answer what makes a body fall, he was able to describe exactly *how* it falls, and with mathematical precision: at an acceleration of thirty-two feet per second.[22] And such mathematical formulations could be verified through experimentation. That level of precision is something that the ancients had been unable to offer, and it is what ultimately provided the impetus to the wide scale acceptance of the method. As Gordon Clark notes:

> The rapid advances of the new scientific method—not only on the grand celestial scale of Kepler and Newton, but as well in the innumerable details of the pendulum and small machines, atmospheric pressure, physical optics, thermodynamics, and the atomic theory—confirmed the limitless power of mathematics and the fruitfulness of quantitative measurement. Therefore, the thesis was inevitable that all natural phenomena are amenable to this method.[23]

Newton's astonishing success transcended the bounds of natural philosophy and profoundly impacted the educational system within Europe. The assumption that Newton's method could be applied to other fields lead to a re-prioritization of subjects to be studied. Traditional subjects such as metaphysics and theology took a backseat to the likes of physics, biology, astronomy, and mathematics. Bacon's *Advancement of Learning* had called for a restructuring of education and was based on the assumption of a universal method. However, he was unable to provide the tangible results to convince others of the efficacy of that method. Newton's success provided the impetus for such a change to occur. And while an Englishman ultimately succeeded in proving the effectiveness of such a method, the French became the torchbearers of Bacon's vision.

The Encyclopedists

Inspired by the "English philosophy" of Bacon, Newton, and Locke, the Encyclopedists established themselves as an intellectual and political force within France.[24] They consisted of prominent members of the *philosophes*, including D'Alembert, Diderot, Rousseau, and Condillac. D'Alembert's *Preliminary Discourse to the Encyclopedia* is perhaps *the* representative document of the Enlightenment.[25] It was a declaration of principles that "represented the views of a party of men of letters who were convinced that through their combined efforts they could substantially contribute to the progress of humanity."[26] It served as the kind of cooperation that Bacon had called for in his *Great Instauration*. The *Encyclopedia* was "to contain nothing less than the basic facts and the basic principles of all knowledge."[27] It represented "a new *summa* of all the branches of knowledge in the light of the major discoveries that had been made in the past one hundred years—a synthesis based upon secular and naturalistic principles rather than upon a traditional theological teleology."[28] The foundation of such an effort was to be found in a unifying *method*, one borrowed from Newton. The method and system envisaged "was designed to organize all our valid information and at the same time to facilitate the discovery of more facts and principles that would be *useful* to humanity."[29]

The influence of Bacon is apparent throughout the *Discourse*. D'Alembert does not hesitate to acknowledge his indebtedness to Bacon:

> One would be tempted to regard him as the greatest, the most universal, and the most eloquent of the philosophers, considering his sound and broad views, the multitude of objects to which his mind turned itself, and the boldness of his style, which everywhere joined the most sublime images with the most rigorous precision. Born in the depths of the most profound night, Bacon was aware that philosophy did not yet exist.[30]

Some of the key features of D'Alembert's philosophy were taken directly from Bacon: "for Bacon's intellectual utilitarianism, for his doctrine that knowledge is power, for his vision of mutual cooperation of all scholars in the progress of knowledge, and for his rejection of scholastic teleology, D'Alembert saluted him as a kindred spirit and great pioneer."[31] While D'Alembert ultimately rejects Bacon's "extreme" inductive method, the spirit of Bacon's work is echoed throughout the *Discourse* and the *Encyclopedia*.

While the goals of the Encyclopedists closely mirrored those of Bacon, the methods employed were primarily Newtonian. D'Alembert sought to unify the "rationalism" of Descartes with the "empiricism" of Locke and Bacon and used Newton's method to do so.[32] As Schwab notes:

> He put forth as the supreme method of achieving truth a combination of rationalism and empiricism closely paralleling the scientific method of the great Newton, which started with sense evidences, or "facts," as the foundation for the discovery of all laws or principles, and which undertook to analyze each object or problem of the physico-mathematical sciences into its simplest elements and thereby to discover the essential principles behind it. The assumptions of the unity, simplicity, and continuity of phenomena in no way prevent the unprejudiced study of the facts presented by the senses.[33]

D'Alembert also shared Newton's misgivings about mixing metaphysics with science. Indemonstrable, *a priori* assumptions, such as those offered by Descartes and Leibniz, were rejected outright as having no place in true science.[34] However, while Newton did address such questions and realized their importance, D'Alembert eschews them completely and seeks to apply Newton's method to the whole range of human knowledge; something Newton never intended.

The application of Newton's method to other fields of knowledge follows Bacon's vision of a unifying method, but the prestige associated with Newton had drastically enhanced the acceptance of such an idea. One such field that was particularly influenced by the new science and its unifying method was ethics. D'Alembert outlines a purely materialistic ethical system, one divorced from any metaphysical foundations. The worldview presented closely mirrors that presented by Voegelin in his analysis of Newton's *Principia*. The universe now simply consisted of a "huge machinery of dead matter, running its course by the inexorable laws of Newton's mechanics."[35] Such a universe has no need of God, and thus man becomes the focal point of the universe. Under such conditions, self-preservation in the Hobbesian sense becomes paramount.[36] Man can shape his own destiny, but needs security in order to do so. D'Alembert essentially reduces ethics to the pursuit of pleasure and the avoidance of pain. The root of evil lies not in man's fallen nature, but in the unequal distribution of wealth; something that can be corrected by a reorganization of society.[37] His failure to recognize God as the source of order for his ethical system drew the ire of orthodox Christians, but generally reflected the outlook of Enlightenment thought. Moral conscience was shaped, not by the divine, or by an unwavering natural law, but by the "cumulative historical experience of mankind."[38] D'Alembert and his counterparts could "turn to history and isolate the essentials of human moral nature, and thus the essentials of moral and political science, through the accumulation and comparison of a vast number of cases of human moral behavior."[39]

The turn to history for guidance marks another significant development to arise from the French Enlightenment. This is not to say that previous thinkers had not utilized history as a guide to action,[40] but the *movement* of mundane history and the idea of *progress* is something that can directly be attributed to the Encyclopedists and *philosophes*. The discovery of *the* unifying method meant that progress was inevitable; man had finally found the key to controlling nature and the old impediments to knowledge, such as theology and metaphysics, would wither away as enlightenment spread. A practical implication of such a turn was the emergence of the idea of an "authoritative present." As Voegelin succinctly notes, "When the intellectual and spiritual sources of order in human and social life dry up, there is not much left as a source of order except the historically factual situation."[41] But that order requires a source of legitimacy, and the idea of progress supplies it:

> The idea of progress through several phases of history, supported by an array of materials which show the increase in value through the successive phases, furnished the basis for this first necessary assumption. The idea of progress, however, creates legitimacy for the present only insofar as it evokes its superiority over the past. Hence, typically in the doctrine, a second idea recurs which is destined to protect the present against invalidation by the future. . . . [T]he present is considered the last phase of human history; no situation of the future ought to differ in substance from the situation envisaged as the desirable present.[42]

In other words, while the future may provide a more perfect realization of the principles set forth by the likes of D'Alembert and Diderot, they will forever be remembered as having set humanity on the path to progress.

Along with the protection against invalidation by the future, the present holds vast superiority over the past. This is due to the fact that man only began to find the true path with the light of modern science, and more specifically with the philosophy of Francis Bacon. Just as Bacon had lamented the fruitlessness of classical and medieval philosophy, the Encyclopedists found little value in consulting Christian or classical sources for guidance. In fact, they sought to justify the exclusion of such sources from education altogether. This is due to several factors, but the most important lies in the rejection of the *bios theoretikos*, or contemplative life. Following Bacon, D'Alembert considered true knowledge to be *useful* knowledge. To have relevance in his system, knowledge must lead to some fulfillment of human needs, and due to the mechanistic worldview, this primarily meant *material* needs. Voegelin notes:

D'Alembert, apparently, has never experienced either the desire for, or the obligation to pursue, that life of contemplation, which Aristotle describes as the *bios theoretikos*. He ignores the fact, or does not know, that the life of man does not exhaust its meaning on the level of utilitarian desires and needs, and that the life of contemplation, resulting in the understanding of man himself and of his place in the universe, is a fundamental spiritual obligation quite independent of its contribution to "useful" activities.[43]

In short, the reductionism inherent in D'Alembert's philosophical anthropology has consequences that cannot be expressed in utilitarian terms. The *spiritual* dimension of man, a central feature of both classical and especially Christian philosophy, is essentially ignored by D'Alembert. And the advancement of science, while certainly beneficial in regards to material well-being, does nothing to fill the spiritual void left by the abandonment of classical philosophy and Christianity.

D'Alembert does realize that religion is a necessary feature for order within a society, but reduces it to a few basic tenets: "One would do a great service to mankind if one could make men forget the dogmas; if one would simply preach them a God Who rewards and punishes and Who frowns on superstition, Who detests intolerance and expects no other cult of man than mutual love and support."[44] Moreover, "some truths to be believed, a small number of precepts to be practiced; such are the essentials to which Revealed religion is reduced."[45] Man, not God, sits at the center of D'Alembert's philosophy. God is nothing more than "an abstraction reasoned naturally from the senses" and there is no mention of the need for salvation, of miracles, or the importance of Scripture.[46] D'Alembert's call for a "reasonable" religion fits well into the scheme first outlined by Bacon. And like Bacon, he presents it as a return to primitive Christianity, not as a radical break from it. But as Voegelin points out:

This reform does not imply a *renovatio evangelica*. It does not have its source in a mystical experience and it implies no more than rationalist purification of Christian symbols, including the divinity of Christ, so that in the end Jesus appears as a "sort of philosopher" who counsels mutual love and support without any intelligible authority or foundation for such counsels. We can observe here in formation the highly important merger of spiritual obscurantism with the apprehension that a religious substitute for Christianity might be necessary, and that the substitute would even have to include a cult.[47]

D'Alembert fails to provide the basis for such a cult, but his philosophy prefigures the cults of Saint-Simon and Comte.[48]

Before moving on to those towering figures, it is important to note one final feature of D'Alembert's *Discourse* and the *Encyclopedia*: the emphasis on technology. The *Encyclopedia* was designed to appeal to the general population. This is in part, because the transformation of society would require their political support. But perhaps more importantly, D'Alembert and Diderot recognized the role that technology could play in that transformation. This is evident by the inclusion of the manual arts and trades in the *Encyclopedia*. The tradesman and artisan were not to be seen as "beneath" the realm of philosophy, but had to be included as integral parts of any movement to transform society.[49] The emphasis on utility, and rejection of the *bios theoretikos* led to the effusive praise of the mechanical arts. And like Bacon, D'Alembert thought it was important to honor those who provided humanity with *useful* inventions. D'Alembert notes:

> The contempt in which the mechanical arts are held seems to have affected to some degree even their inventors. The names of these benefactors of humankind are almost all unknown, whereas the history of its destroyers, that is to say, of the conquerors, is known to everyone. However, it is perhaps in the artisan that one must

seek the most admirable evidences of the sagacity, the patience, and the resources of the mind. . . . Why are not those to whom we owe the fusee, the escapement, the repeating-works of watches equally esteemed with those who have worked successively to perfect algebra?[50]

The implication of course is that if inventors were to be honored properly, society would benefit. And moreover, those who did *not* provide society with *useful* knowledge would be condemned to obscurity: "all occupations with purely speculative subjects should be excluded from a healthy state as profitless pursuits."[51]

This type of attitude led to a reorganization of the educational system in France, and perhaps no institution reflected that change as well as the *Ecole Polytechnique*. Hayek calls it the "source of scientistic hubris" and explains the drastic change in curriculum:

In conformity with the ruling spirit and by an overviolent reaction against the older schools, the teaching in the new institutions was for some years confined almost exclusively to the scientific subjects. Not only the ancient languages were reduced to a minimum and in practice almost entirely neglected, even the instruction in literature, grammar and history was very inferior, and moral and religious instruction, of course, completely absent.[52]

Saint-Simon, reflecting upon the changes in the educational system, notes, "In the not so distant days, if one wanted to know whether a person had received a distinguished education, one asked: Does he know his Greek and Latin authors well? Today one asks: Is he good at mathematics?"[53] This was exactly the type of change that D'Alembert and his fellow collaborators had hoped for when composing the *Encyclopedia* a mere sixty years earlier. And it was further confirmation that Newton's method had gained widespread acceptance, not only in intellectual circles, but also among the general population. Yet, the new method still lacked the spiritual authority necessary to truly convince others to act. Saint-Simon took that to be his task and it is to his work that I must now briefly turn.

Saint-Simon

Henri de Saint-Simon (1760–1825), writing in the wake of the French Revolution, was hoping to re-unify society under the auspices of science. He was greatly influenced by the *Encyclopedia* and *philosophes*, but drew the most inspiration from the work of Isaac Newton. He praised the "positive sciences" and thought that the science of man should follow suit. Since society was still governed by those who do not understand the general laws that rule the universe:

> It is necessary that the physiologists chase from their company the philosophers, moralists, and metaphysicians just as the astronomers have chased out the astrologers and the chemists have chased out the alchemists. . . . We are organized bodies; and it is by regarding our social relationships as physiological phenomena that I have conceived the project which I present to you.[54]

Hayek points out that the "physiologists themselves are not yet quite scientific enough" since "they have yet to discover how their science can reach the perfection of astronomy by basing itself on the single law to which God has subjected the universe, the law of universal gravitation."[55] That task is to be carried out by the Council of Newton, "by exercising its spiritual power to *make* people understand this law."[56]

Saint-Simon's Council of Newton would consist of twenty-one scholars and be presided over by the mathematician who had received the most votes.[57] The Council would serve as "the representatives of God on earth, who would deprive the pope, cardinals, the bishops, and the priests of their office because they do not understand the divine science which God has entrusted to them and which some day will again turn earth into paradise."[58] And what happens to those who do not go along with Saint-Simon's plan? "Anybody who does not obey the orders will be treated by the others as a quadruped."[59] This is because "the law of human progress guides and dominates all; men are only its instruments."[60]

To say that Saint-Simon rejected the tenets of liberalism would be an understatement. Diversity, especially in ideologies, would only lead to civil strife as in the French Revolution. Saint-Simon notes, "The vague and metaphysical idea of liberty . . . impedes the action of the masses on the individual . . . and is contrary to the development of civilization and to the organization of a well-ordered system."[61] Individual liberty must give way to collective necessity, especially if the desired ideological uniformity is to be achieved. Saint-Simon recognized the utility of religion in bringing about this uniformity, but just as the Encyclopedists had noted, the old forms of Christianity would no longer be sufficient. All religion had to be brought in line with "positive science" to be considered legitimate. While he generally rejected the teachings of Christian doctrine, he found worth in the institutional structure of the Church. The scientists were to organize themselves along the lines of the Catholic clergy and direct the education of the masses. Furthermore, they should "elect a scientific pope, employ excommunication for crimes against the ideology, and institute a Newtonian form of baptism."[62] Thus, we can see the far-reaching implications of the scientistic attitude. It is not merely enough to promote science as a good for society; instead, it becomes *the* good and political, economic, and even religious institutions must reflect its principles.

Saint-Simon, reflecting on his work a few months before his death, provides us with a succinct summary of his efforts:

> Like all the world, I wanted to systematize the philosophy of God. I wanted to descend successively from the phenomena of the universe to the phenomena of the solar system, further on to the terrestrial phenomena, and finally to the study of the species considered as a subdivision of sublunar phenomena; from this study I

wanted to derive the laws of social organization, the original and essential object of my enquiry.[63]

As Voegelin notes, "Saint-Simon confesses a dream that was inspired by Newtonian physics, the dream of expanding the type of science that had evolved in mathematical physics to embrace systematically the whole world, including human society."[64] Yet, as Saint-Simon laments, that dream was never able to come to fruition because he found it impossible to extend Newton's fundamental law to other fields.

Saint-Simon's failure to find a unifying law led him to change his approach:

> He will no longer pursue this aim (the exploration of the social organization) by searching for a social law of gravitation, rather, he will pursue it by the development of blueprints for a society in the age of Newtonian science, technology, and industry. Scientists and philosophers have the "positive" function of developing the systematic body of knowledge that will guarantee human domination over nature. The social organization will abolish the old domination of man over man and replace it by a government of scientists, engineers and industrialists who secure and increase the domination of man over nature for the benefits of society at large.[65]

In other words, government is dominated by technocrats and becomes administrative and managerial in nature. This of course partially prefigures what would later come with Engels and Marx, but it also represents a continuation of Bacon's program. Solomon's House was composed of scientists who would ultimately decide political issues and Saint-Simon's Council of Newton serves a strikingly similar purpose. More importantly, those institutions would also provide *spiritual* guidance.

Although Saint-Simon's project represents the scientistic attitude and carries on the tradition started by Bacon and continued by the Encyclopedists, his work is less than systematic. And while he often espoused the great benefits of science and of a scientific education, he was not formally trained in any particular science and offered no scientific discoveries. He can be generously characterized as a partisan for science. He saw positive science as a way to overcome the factions that resulted from the French Revolution. He wanted to unify society using positive science, but simply could not devise a coherent program to accomplish that task. In spite of his own inability to do so, Saint-Simon had an appreciable influence on several younger scholars. The most notable of these, Auguste Comte, was able to systematize the philosophy of Saint-Simon. Comte was much more versed in philosophical matters than his mentor and had a much better command of history. More importantly, he was able to express his ideas coherently through his writings. He sought to "positivize" the study of man and extend Bacon's method beyond natural philosophy to social philosophy. While others had spoken of the need for a positive science of man, Comte was the first to explicitly sketch such a science. Drawing from the vision of Bacon, the mathematical precision of Newton, the ambition of the *philosophes*, and the spiritual impetus of Saint-Simon, Comte outlines a unified, totalizing positive system that encompasses *all* aspects of human life. His work is representative of the drastic changes accompanying French, and moreover European, society. It represents an intellectual *and* spiritual revolution. And thus it is to his work that I now must turn.

Chapter 4
Auguste Comte's Scientistic Politics

"In the name of the past and the future, the theoretical servants and the practical servants of Humanity assume befittingly the general leadership of the affairs of the earth in order to construct, at last, the true providence, moral, intellectual and material; they irrevocably exclude from political supremacy all the various slaves of God, Catholics, Protestants, or Deists, since they are retrogrades as well as pertubators."[1]—August Comte

Auguste Comte (1798–1857) stands as one of the most influential figures of the nineteenth century. As part of the generation to directly follow the French Revolution, Comte's primary concern was to restore political order and to find a balance between the retrograde and revolutionary forces within French society. Comte thought a return to pre-Revolutionary principles to be impossible, yet only saw destruction in the principles espoused by the Revolutionaries. Furthermore, any solution that failed to take into account *all* of society's primary institutions would be insufficient. The political situation called for a complete reordering of society and its institutions, and Comte's system provided the blueprint. Comte sought to reorder every major institution from the family to business to the Church and ultimately to the regime itself.[2] Comte saw himself as the "Aristotle of Positivism" and expected his work to have the same type of impact that Aristotle's had, especially in Europe where the Aristotelians had been so influential in the Middle Ages.[3] Yet the disintegration of Christianity's influence and the rise of immanent materialism left more than just an intellectual void; the *spiritual* void created by the "new" worldview was immense. Thus, in addition to being the "Aristotle of Positivism" and establishing a new way to study man and understand society, Comte also had to mirror St. Paul and establish a new spiritual order. He did so by founding the Religion of Humanity as a replacement for Christianity. Comte's synthesis of a systematization of science and a "humanistic, demystified" religion is what sets him apart from his predecessors.[4] While some have tried to separate those two elements, to do so is to miss the spirit of Comte's work and to artificially implement a division that he rejected.[5] Viewed in that light, I will explore *both* elements of his work and demonstrate the continuity inherent in his philosophy. Moreover, I will demonstrate its *scientistic* character and to borrow from Eric Voegelin, I will show how Comte plays the role of "intramundane eschatologist."[6]

Before delving into the works themselves, it is important to note some of the major influences on Comte, as well as his place within the history of political ideas. Comte was born into a devout Roman Catholic family with royalist leanings. By the age of 14, he had become an atheist and placed himself firmly in the camp of the republicans.[7] He was initially denied entrance into the esteemed *Ecole Polytechnique* because of his young age, but was eventually allowed admission. However, just two years into his tenure, he was among the many expelled for "radicalism."[8] Shortly thereafter, Comte entered into one of the most important relationships of his life. He became secretary for the renowned social theorist, Henri de Saint-Simon.[9] Comte spent seven years under the tutelage of Saint-Simon, and in spite of an acrimonious break, Saint-Simon's influence remained with Comte throughout his career. Saint-Simon's call for a "positive" science of man and for a reorganization of society based on scientific principles greatly appealed to Comte, as did his philosophy of history. Although these ideas resonated with Comte, he "grew increasingly unhappy about Saint-Simon's weakness in technical rigor and what Comte saw as a scandalous disregard for scientific-empirical modes of argument."[10] Shortly after the break with Saint-Simon, Comte set about to correct that deficiency by establishing a new foundation for philosophy. Unlike the works of his predecessor, the *Cours de philosophie positive* would represent a systematic, scientific exposition of positive philosophy. Perhaps more importantly, it would serve as the capstone of a movement started two centuries earlier by Francis Bacon. Before continuing to Comte's works, it is important to note the Baconian influence on Comte's project.

Bacon and Comte

In many ways, Comte's project serves as the continuation of Bacon's *Great Instauration.* As Andrew Wernick notes:

> The guiding aim of this current was to develop a grand synthesis of scientific knowledge through systematically mapping its results and principles. This in turn would provide a basis both for a naturalistic understanding of humanity's place in the cosmos, and for forging an intellectual instrument for extending human control. In historical terms, the project presented itself as a correction of Aristotle—particularly the Aristotle of the Scholastics—in light of the rise of the natural sciences. If the initial target was Aristotelianism, however, the principal thought-opponent was Plato, and by extension all aprioristic, idealist, in short "metaphysical" forms of reason. Comte's contribution was to apply this critique to the rationalist political and moral theory of the *philosophes* so that, by means of a real "science of Man," the Baconian matrix could be fully positivized as the subject and object of its own gaze.[11]

Bacon had been the first to call for a radical reorganization of both education and society based on the new science. Bacon's experimental method had opened the door to a new way of approaching the world. Moreover, his vision of a scientific utopia, presented in the *New Atlantis,* had provided a goal to aspire to. While Bacon provided little, if anything, in regards to actual scientific discoveries, his advocacy of the experimental method had a profound influence. Newton's emergence might not have been possible if not for Bacon's efforts. Bacon's influence was readily apparent in the creation of the Royal Society and in the writings of Comte's French predecessors, particularly the Encyclopedists. It is instructive to note some of the key themes that Comte takes from Bacon, as well as a few points on which they diverge.

Bacon and Comte both thought that medieval and classical philosophy had not led to the type of progress that humanity was capable of achieving. The impoverishment of natural philosophy was a direct result of these flawed modes of thought, and society had suffered as a result. The solution was to provide a new way of thinking; one that would lead to tangible results and admits of demonstration. But before the new philosophy could take hold, an explanation had to be given as to why the previous modes were inadequate. Bacon's primary argument is that those modes provided humanity with very little *useful* knowledge. Metaphysical speculations were fruitless; the questions asked were never answered adequately as the different camps would simply stick to their preferred dogmas. And even if the questions had been answered, they left little room for a direct impact on improving man's condition. Comte also argues that the previous modes of philosophizing were fruitless, but he further "strengthens" his position through his law of the three states. Bacon had argued against the theological and metaphysical modes of thought, but simply relied on reason, and hope, to make his case. Comte appeals to an invariable law and to the authority of History itself. He does not simply suggest a new mode of philosophizing: he outlines a system that *must* be accepted.

Once the previous modes of thought have been refuted and a case has been made for the adoption of the new science, the question of implementation becomes paramount. Both Bacon and Comte outline a dramatic restructuring of education. Bacon does so primarily through the *Advancement of Learning*, while Comte utilizes the *Cours de positive philosophie*. Bacon's restructuring significantly downgrades the formerly prominent roles of theology and metaphysics and promotes the natural sciences. Comte's scheme seeks to eradicate the theological and metaphysical philosophies completely and focuses on the six positive sciences: mathematics, astronomy, physics, chemistry, biology, and sociology. Both outline a new *method* and claim that it can be applied universally to all forms of knowledge. In fact, the method used is one of the two main criteria, the other being utility, that determine if something constitutes true knowledge or not.

Bacon and Comte both realized that educational reform was a primary requisite for the transformation of society, but it alone would not be sufficient for a successful social reorganization. A *spiritual* change had to accompany the intellectual or there would be no sufficient impetus for action. This is an area where Bacon and Comte diverge, but perhaps more in method than in spirit. Bacon's society was firmly entrenched in Christian principles, and the Church held significant power. The new science was just emerging and tangible benefits were not readily apparent. The great light of Newton had not yet arrived so Bacon had to be careful not to disturb the prevailing powers. He was cautious in presenting his political ideas and did so primarily through cryptic aphorisms, as in the *Essays*, and through myth. And even in the *New Atlantis*, an overtly fictional work, Bacon was sure to include Christianity in his presentation. His proposed solution, a society based on scientific principles, had to *appear* friendly to Christianity, both for the acceptance of the primarily Christian population and more importantly, the regime. On the other hand, Comte did not have to provide much evidence as to the efficacy of the new science, and he certainly did not have to mince his words in regards to religion. Christianity was on the decline as Enlightenment principles had taken firm hold, especially within French society. Thus, an attack on Christianity was nothing new and Comte did not have to worry about the kind of negative consequences that Bacon did in regards to his stance on religion. As a result of this, Comte did not have to conceal his intentions through myth or aphorism, and he could explicitly outline the precepts of a new religion.

As previously noted, the emphasis on *method* was a common feature of both Comte and Bacon's projects, but they also shared a view of the ends sought through that method. Both wanted to improve man's condition and to reorganize the social structure. Thus, *utility* became of paramount importance. The method was intended to produce useful knowledge.[12] Bacon makes this clear by promoting experiments of "fruit" and praising inventors as demi-gods within *New Atlantis*.[13] Likewise, Comte praises the positive philosophy for its ability to bring about tangible benefits to society and honors inventors throughout his Positivist Calendar.[14]

The corollary to the emphasis on utility was the diminishment of the *bios theoretikos,* or contemplative life. Bacon places blame on the ancients, particularly Plato and Aristotle, for the feeble state of natural philosophy.[15] He characterizes the Greeks as engaging in a "kind of childish stage of science . . . too ready to talk, but too weak and immature to produce anything."[16] Likewise, Comte likens the theological and metaphysical stages of philosophy to childhood and immaturity.[17] Aside from their flawed methodology, these systems produced little in the way of tangible benefits to society, the primary criterion for success in the eyes of both Bacon and Comte. The lack of practical results can largely be attributed to the questions asked by the ancients. They were either unanswerable (i.e., questions of first principles) or "fruitless."[18] While they disagreed on the role of such questions, it is clear that both Bacon and Comte left little room for them within their own projects.[19] Bacon's project was aimed to help "relieve man's estate" and science was the means through which that task could be accomplished. Similarly, Comte sought to reorganize society on positive science. Societal needs drive the project and a true science is one that fulfills these needs.[20]

In spite of their similarities, Comte faced a major challenge that Bacon had not been forced to deal with. Bacon had based his project on two primary assumptions: that humans were entitled to, and capable of, dominion over nature and that the fruits of that dominion should benefit all.[21] Bacon took care to fit his project into the Christian framework and appealed to charity to do so. Contrary to the notion that the advancement of knowledge "has somewhat of the serpent, and puffeth up," Bacon argued that the "proud knowledge of good and evil" had led to the fall.[22] God had intended man's dominion over nature, and the new science provided the means to do so. As long as that dominion of nature was used to the benefit of mankind, it fit comfortably in the confines of Christian principles. Without such a principle underlying the project however, the domination of man over nature could easily drift into the domination of man over man.[23] In the two centuries that elapsed between Bacon and Comte, the influence of Christianity had waned. Thus, Comte had to find a new moral foundation. As Wernick notes:

> With the declining credibility of revealed religion, techno-progressive thinking faced a dilemma. Either the moral grounding of a world-mastering science would have to be left unsecured—letting the enterprise float, in principle and *de facto,* among the powers of the world, or some non-supernatural way would have to be found to legitimate the philanthropic ethic which had been shored up, in the old religion, by the word of God.[24]

The Religion of Humanity was intended to do just that as Comte looked to not only replace the Christian foundation that had been displaced, but to improve upon it. The new religion would be based on positive science, the triumph of which would be guaranteed by History. Given its scientific foundation, it would be "demonstrable" and thus truly universal. Now I will turn to Comte's works and examine both the scientific foundation, established in the *Cours de positive philosophie*, and the new religion itself.

Cours de positive philosophie

While I cannot presently explore every detail in the six-volume work that constitutes the *Cours*, it is important to note the *purpose* of the work and to explore its central arguments. In light of the great advances in the natural sciences, especially since Newton's time, Comte lamented the lack of progress in the social sciences. The reason for such stagnation was that the *method* that had been successfully utilized in fields such as physics, astronomy, and chemistry had not been applied to the social. The science of man had yet to be "positivized" and Comte set out to continue the "philosophical impulse of Bacon and Descartes" by establishing a "social physics."[25]

Before outlining the characteristics of such a science, Comte had to first explain why a change was necessary and why the previous modes of exploring social questions were deficient. To do so, he posited a "great fundamental law, to which the mind is subjected by an invariable necessity."[26] The law of three states (or stages) can be "demonstrated both by reasoned proofs furnished by a knowledge of our mental organization, and by historical verification due to an attentive study of the past."[27] Comte continues:

> This law consists in the fact that each of our principal conceptions, each branch of our knowledge, passes in succession through three different theoretical states: the theological or fictitious state, the metaphysical or abstract state, and the scientific or positive state. In other words, the human mind—by its very nature—makes use successively in each of its researches of three methods of philosophizing, whose characters are essentially different and even radically opposed to each other. We have first the theological method, then the metaphysical, and finally the positive method. Hence, there are three kinds of philosophy or general systems of conceptions on the aggregate of phenomena, which are mutually exclusive of each other. The first is the necessary starting point of human intelligence; the third represents its fixed and definitive state; the second is destined to serve only as a transitional method.[28]

It should be noted that the law of the three states is not meant to describe how things *should* progress, but rather how they *must* progress. The movement of history dictates that positivism will triumph and Comte is merely the instrument that will bring it about.[29]

The first state, the theological, is one in which the "human mind directs its researches mainly towards the inner nature of beings, and toward the first and final causes of all the phenomena that it observes—in a word, toward absolute knowledge. It therefore represents these phenomena as being produced by the direct and continuous action of more or less numerous supernatural agents, whose arbitrary intervention explains all the apparent anomalies of the universe."[30] It corresponds to theocracy and includes the movement from fetishism to polytheism and finally to monotheism.[31] It represents an "embryonic" form of knowledge and is characterized by the search "for the hidden nature of things" as the mind tries to find out *why* things happen as they do with the ultimate explanation coming from a divinity created in man's image.[32]

The second state, the metaphysical, represents a transitory stage in which the theological state is modified through the replacement of supernatural agents by "abstract forces, real entities or personified abstractions, inherent in the different beings of the world."[33] The mind "still wants to know the 'why' of phenomena, but it accounts for what happens differently, by creating secular or natural divinities, as it were, which man now holds responsible for the observed facts: forces, qualities, powers, properties and other such constructs."[34] As Kolakowski notes, "the metaphysical state undergoes a development similar to that of the preceding state, culminating in a kind of secular monotheism, which compresses the multiplicity of occult powers into the single over-all concept of 'nature,' regarded as capable of accounting for all the facts."[35]

It is not until the third state, the positive, that the human mind finally reaches its potential:

> Finally, in the positive state, the mind, recognizing the impossibility of obtaining absolute truth, gives up the search after the origin and hidden causes of the universe and a knowledge of the final causes of phenomena. It endeavours now only to discover, by a well-combined use of reasoning and observation, the actual laws of phenomena—that is to say, their invariable relations of succession and likeness. The explanation of facts, thus

reduced to its real terms, consists henceforth only in the connection established between different particular phenomena and some general facts, the number of which the progress of science tends more and more to diminish.[36]

In other words, positive philosophy rejects questions of formal and final causes, and concerns itself instead with material and efficient causes. Furthermore, it seeks to reduce those causes to the fewest number of laws possible in an attempt to unify knowledge. As Kolakowski points out, "it asks *how* phenomena arise and what course they take; it collects facts and is ready to submit to facts; it does not permit deductive thinking to be carried too far and subjects it to the continuous control of 'objective' facts."[37] The positive phase is characterized by *certainty* and seeks to discover invariable universal laws.[38] Observation, experiment, and calculation are the means to finding such laws. Any question that cannot be explored through such means is deemed unimportant or superfluous in the positive state of philosophy, and *utility* is the criterion for what constitutes knowledge. The ultimate goal is the improvement of man's condition (not individually, but collectively) and anything that does not contribute to that end is useless.

While the law of three states applies to all fields of knowledge, it progresses at different rates. This is due to two factors: complexity and generality. The least complex and most general science, Mathematics, was the first to become positive. This is because it deals with quantity and every measurable relationship between phenomena.[39] Astronomy was the second field to become positivized. Its range is more limited than that of mathematics (i.e., it is less generalizable), but it is "richer for bringing within the range of science a further feature: force."[40] Physics is next in the ordering of sciences and introduces qualitative features such as heat and light. Chemistry follows suite with "qualitatively differentiated substances."[41] Biology constitutes the final science to be positivized before Comte's founding of social physics. It is characterized by "an even greater number of qualities" than chemistry, but with a narrower range of subject matter and investigates organic structures.[42] Finally, social physics, or sociology, represents the last field and crowning achievement of science. Its object, human beings and their society, is the most complex and least universal of all the subjects to be studied.[43] Unlike the preceding sciences, social physics had yet to be constituted, and Comte's primary task was to do just that.

As Kolakowski notes, the ordering of sciences is "at once logical, historical, and pedagogical. That it is logical is apparent from the fact that it is based on mutually consistent principles. We realize it is historical when we observe that each science reached the positive stage at a different point in time." And it is pedagogical in the sense that Comte intends for the "sciences to be taught in the order of their development, so that they may form a coherent system in the student's mind."[44] Comte not only recommends such a program; he finds it *essential* to the acquisition of knowledge. According to this assumption, one cannot learn astronomy without first learning mathematics or chemistry without first having studied mathematics, astronomy, and physics. Thus, social physics would require the most strenuous work since it not only contains the most complex subject, but also the study of mathematics, astronomy, physics, chemistry, and biology as prerequisites.

It is important to note a few other consequences of Comte's ordering of sciences. As Kolakowski points out, Comte "rejects the doctrine that would reduce all disciplines to lower ones." While Comte readily acknowledges that "the more complex sciences presuppose the less complex ones," this does not mean that the "complex phenomena are not subject to irreducible laws of their own."[45] The unity of science therefore is achieved not by reducing the higher to the lower, but by acknowledging the interdependence of the sciences. This interdependence requires those in each discipline not to lose sight of the whole or in other words, not to over-specialize. The unity implied in such an ordering has to do not only with the methods employed, but the ends sought. As mentioned previously, *usefulness* is the fundamental criterion of the value of knowledge and those in each science must be constantly reminded of this fact.[46] *Societal needs* are the driving force of science properly constituted, and this can be obscured if those within each science do not appreciate the interdependence of their discipline on others. As Kolakowski notes:

> If it is not to bog down in fruitless speculation and waste man's intellectual energies, science must continually be reminded of its social tasks, which in the last analysis determine its value. This practical control of knowledge has a historical and social character; it is not effected in the minds of individual scientists, but is effected continuously by the human species as a whole. In the last analysis, then, what science needs is what society needs, namely, the ability to predict events and to influence them practically.[47]

Comte's law of the three states and subsequent ordering of the sciences calls for a dramatic overhaul of both the social and educational institutions of society, but why should we accept such a program? According to Comte, the law of the three states is proven by history. That each science passes through three stages is simply an empirical fact. Furthermore, Comte appeals to individual experience and asks: "does not each of us in contemplating his own history recollect that he has been successively—as regards the most important ideas—a theologian in childhood, a metaphysician in youth, and a natural philosopher in manhood?"[48]

Assuming Comte is right about the historical progression of the sciences, the question still must be asked whether things should continue to progress in such a way. Is such a progression actually good for humanity? To such questions, Comte tersely replies: "whether this is a good or a bad thing matters little; the general fact cannot be denied, and that is sufficient. We may deplore the fact, but we are unable to destroy it; nor consequently, can we neglect it, on pain of giving ourselves up to illusory speculations."[49] In other words, certain questions are prohibited when it comes to progress.

If one is left unsatisfied by the "proofs of the truth of this law furnished by direct observation of the race of the individual,"

Comte offers further theoretical considerations that show its necessity.[50] Comte continues:

> The most important of these considerations arises from the very nature of the subject itself. It consists in the need at every epoch of having some theory to connect the facts, while, on the other hand, it was clearly impossible for the primitive human mind to form theories based on observation. *All competent thinkers agree with Bacon that there can be no real knowledge except that which rests upon observed facts.* This fundamental maxim is evidently indisputable if it is applied, as it ought to be, to the mature state of our intelligence.[51]

Comte does not condemn those in the past for not thinking in such a manner because they were stuck in the primitive stages of knowledge, dictated by the movement of history itself. The theological conceptions were *necessary* for the primitive mind to make sense of the world and the metaphysical conceptions served their purpose of bridging the gap between the theological and positive stages.[52] Yet once the positive stage has been reached, there is no going back. Thus, Comte can assert that "all competent thinkers" must agree with Bacon's theory of knowledge. And that theory of knowledge explicitly denounces metaphysical and theological speculations as being "fruitless."

Once positive philosophy has enabled the "irrevocable break from those fruitless doctrines and provisional methods that were suited only to its first flight," the potential for man to improve his condition is limitless.[53] But that cannot be accomplished until the positive philosophy encompasses *all* classes of phenomena. On Comte's account, the omission of social phenomena from the positive stage is what has ultimately prevented society from being successfully organized. While theological and metaphysical methods have been expunged from investigations of other classes of phenomena, those "discarded methods are, on the contrary, still used exclusively for . . . everything that concerns social phenomena, although their insufficiency in this respect has been fully felt by all good minds, such men being tired of these empty and endless discussions between, e.g., divine right and the sovereignty of the people."[54]

The explicit prohibition of questioning stands as one of the primary features of Comte's system. Questions regarding individual rights and the sovereignty of the people are fundamental to politics, but they are prohibited in Comte's construction. It is not just that Comte refuses to answer them: he bans them outright. There are several reasons for this. One is that such questions cannot be answered by the positive *method*. Yet, one of the primary justifications given for adopting that method is because of the vast success it has created in the natural sciences. Instead of admitting any type of limitation to the method, Comte prefers to simply ban any questions that it cannot answer. Another reason for the prohibition of questions is that Comte has a manic desire for unity. This is something that nearly every scholar of Comte agrees on.[55] He wants unity of method and of ends, and cannot allow for questions that would either lead to different ends or require different methods. Comte fails to consider the possibility that the mere asking of those questions constitutes a fundamental element of human existence. The lack of a consensus in the answers to such questions should not be taken to mean that the questions should not be asked or that they hold no value. To the contrary, it can be argued that the *most* meaningful questions in human life are ultimately unanswerable.

That the prohibition of questioning can lead to deleterious effects is something that Comte does not seem to fully appreciate. As Voegelin notes:

> Today, under the pressure of totalitarian terror, we are perhaps inclined to think primarily of the physical forms of opposition. But they are not the most successful. The opposition becomes truly radical and dangerous only when philosophical questioning is itself called into question, when *doxa* takes on the appearance of philosophy,

when it arrogates itself to the name of science and pro-
hibits science as nonscience. Only if this prohibition can
be made socially effective will the point have been
reached where *ratio* can no longer operate as a remedy
for spiritual disorder.[56]

Voegelin points to Comte as a prime example of a thinker who explicitly employs such a prohibition. According to Voegelin, Comte
knows his system cannot stand up under critical examination and
therefore bans questions about its premises. The "conscious, deliberate, and painstakingly elaborated obstruction of *ratio* constitutes
a new phenomenon."[57] Those who ask questions about "the nature, calling, and destiny of man may be temporarily ignored," but
later on, once positivism has taken hold in society, such people
"will have to be silenced by appropriate measures."[58] Here we can
recall Saint-Simon's solution to those who do not go along with the
plan: "Anybody who does not obey the orders will be treated by
the others as a quadruped."[59]

After establishing the need for a social physics, Comte turns to
the positive philosophy as a whole and seeks to unify the sciences.
As mentioned before, Comte thinks that the end of each science is
ultimately the same: to improve the social condition. Although the
sciences have been positivized, they have not reached their full potential and this is due mainly to a lack of an overall, organizing
structure. Comte thinks that the sciences have become too specialized, and to the extent that this is the case, they have lost sight of
the primary goal: improvement of the human condition. Comte
notes:

> The majority of scientists already confine themselves en-
> tirely to the isolated consideration of a more or less ex-
> tensive section of a particular science, without
> concerning themselves much about the relationship be-
> tween their special work and the general system of pos-
> itive knowledge. Let us hasten to remedy this evil before
> it becomes more serious. Let us take care that the human

mind does not lose its way in a mass of detail. We must not conceal from ourselves that this is the essentially weak side of our system, and that this is the point on which the partisans of theological and metaphysical philosophy may still attack the positive philosophy with some hope of success.[60]

The remedy for this situation is to create "one more great specialty, consisting in the study of general scientific traits. These scientists will determine the exact character of each science and reduce its chief principles to the smallest number of laws or principles within the confines of the positive method."[61] The other scientists, before receiving specialized training, should first receive training in the general principles of positive knowledge. If these two conditions were fulfilled, "the modern organization of the scientific world would then be accomplished, and would be susceptible of indefinite development, while always preserving the same character."[62]

Having established "the general spirit" of positive philosophy, Comte outlines the four fundamental qualities of the positive philosophy. The first is that the study of positive philosophy "furnishes us with the only really rational means of exhibiting the logical laws of the human mind, which have hitherto been sought by methods so ill calculated to reveal them."[63] The method of the positive philosophy will allow us to discover laws that can directly benefit human society. The second result of the positive philosophy is the "general recasting of our educational system."[64] As Comte states:

> Competent judges are already unanimous in recognizing the necessity of replacing our European education, which is still essentially theological, metaphysical, and literary, by a positive education in accordance with the spirit of our time and adapted to the needs of modern civilization.[65]

The problem is that "exclusive specialty" and "too rigid isolation" has inhibited the teaching of the sciences: "An intelligent person who wishes at the present day to study the principal branches of natural philosophy . . . is obliged to study each separate science in the same way and with the same amount of detail as if he wished to become an astronomical or chemical specialist, etc."[66] This prohibits the type of general education that is required for society to progress. Thus, it is "indispensable that the different sciences of which it is composed . . . should first be reduced to what constitutes their essence—that is to their principal methods and most important results."[67] Once this is done, a truly rational, general education can be provided to the masses and scientists alike.

The reorganization of education leads to the third fundamental property: the progress of the different positive sciences. The generalizing effect of the new educational system will help to quell the over-specialization found in each particular science. The artificial divisions created by such specialization hinder progress and prevent the advancements that could be taking place were they not present. As an example of this, Comte points to Descartes and analytical geometry. The genius of Descartes lies in the fact that he simply established "a closer connection between two sciences that had hitherto been regarded from separate standpoints."[68] Likewise, Comte suggests that other sciences will have to interact. For instance:

> It will be indispensable to unite the chemical with the physiological point of view. This is shown by the fact that, even in the opinion of the illustrious chemists who have most powerfully contributed to the formation of this doctrine, the utmost that can be said is that it is always verified in the composition of inorganic bodies; but it is no less constantly at fault in the case of organic compounds, to which up to the present it seems quite impossible to extend the doctrine. . . . [A]n entirely new order of considerations, belonging equally to chemistry and physiology, is evidently necessary in order to decide finally, in some way or other, this great question of natural philosophy.[69]

The fourth fundamental property of the positive philosophy is the most important: it serves as the "only solid basis of the social reorganization that must terminate the crisis in which the most civilized nations have found themselves for so long."[70] Comte bases this assumption on the fact that "the whole social mechanism rests finally on opinions."[71] The great political and moral crisis of existing societies is ultimately due to intellectual anarchy.[72] In fact, he calls such a divergence of opinion the "gravest evil" and notes that its persistence causes an "essentially revolutionary state, in spite of all the political palliatives that may be adopted."[73] The solution of course is to create an agreement on first principles and once obtained, "the appropriate institutions will follow, without giving rise to any grave shock; for the greater part of the disorder will have been already dissipated by the mere fact of agreement."[74] Again, we must recall that the disagreement on first principles is caused by the mutually exclusive, incompatible nature of the theological, metaphysical, and positive philosophies. That empirical fact, on Comte's argument, leaves us with the solution of choosing one, and discarding the other two. But we need not quibble over which path to choose because history has chosen it for us: the positive philosophy *will* prevail.[75]

Before moving on to the second phase of Comte's work, the institution of the Religion of Humanity, it is helpful to recall a few of the main features of the *Cours*, which served as the cornerstone of Comte's overall philosophy. The *Cours* represents an *intellectual* revolution, but with the understanding that it is only the first step in the reorganization of society. The fractured nature of French society, and arguably of Europe as a whole, caused Comte to seek a unifying principle, and he found that in his "discovery" of the law of the three states. He then used that law to justify the founding of social physics and more importantly, the triumph of positive philosophy over the theological and metaphysical philosophies. In doing so, he set forth a progressive theory of history, following in the footsteps of fellow countrymen such as Saint-Simon, Condorcet, Turgot, and D'Alembert. The intellectual victory of positivism was guaranteed by history itself, and questions about the premises of the system were not permitted. In spite of Comte's supreme confidence in the ultimate supremacy of the positive philosophy, he realized that it would be insufficient to enact the drastic social reforms he so desired. The *intellectual* revolution would provide the foundation, but a *spiritual* revolution would be required for the establishment of a positive polity. Thus, I must now turn to the second, and arguably most important, phase of Comte's thought.

Religion of Humanity

What ultimately separates Comte from his predecessors is not his call for a positive science or his articulation of the law of three states; the likes of Saint-Simon, D'Alembert, Diderot, and Turgot had all expressed similar ideas. Comte's founding of the Religion of Humanity and his own "religiousness" is what truly sets him apart:

> Pseudo-prophetic charisma is the strength of Comte, and while his church was not much of a success, his religious enthusiasm was strong enough to endow a body of ideas, although of dubious scientific value, with the glow of a revelation on whose acceptance depends the salvation of mankind. Comte has not added much as a thinker to the complex of Positivist ideas; he has added to them in his capacity as a religious founder by shifting them to the level of a dogmatic religion.[76]

The new worldview, first brought about by Newton's *Principia*, had left a spiritual void and Comte recognized the necessity of filling that void. His Religion of Humanity was designed to replace Christianity, and just as the new science had displaced the old through empirical verification, so too would the new religion take its rightful place. In line with his positivist philosophy, Comte offered a "demonstrable" religion that would supplant the "revealed" religion of Christianity. After serving as the "Aristotle of

Positivism," Comte took on the mantle of St. Paul in the founding of his new religion. I will now outline the fundamentals of this religion, as well as the primary motivations behind its implementation.

Comte's emphasis on unity in the *Cours* extends to his *System of Positive Polity* and the Religion of Humanity. The two dominant political forces in France, the "retrogrades" and "progressives," had failed to establish a satisfactory political order. As Wernick notes, the retrograde party consisted mainly of "royalist counter-revolutionaries attached to the illusory absolutisms of Catholic dogma."[77] Opposing them were those who had triumphed in the first stage of the French Revolution and "who subscribed to a metaphysical belief that individuals could spin workable utopias out of their individual brains. Divided by a one-sided attachment either to reason and progress, or to faith and order, both camps were equally doctrinaire and equally incapable of thinking the reconstruction of the shattered social order in line with its actual laws and requirements."[78] Comte's proposed solution then was to synthesize the two primary elements of each: order and progress. Comte's establishment of sociology would serve as the scientific basis on which to enact this solution and the Religion of Humanity would serve as the primary ordering source within society.

Although Comte had a disdain for the theological principles of the retrograde party, he had an appreciation for the order that religious uniformity could bring about. Comte especially found Catholic organizational principles appealing, although this admiration did not extend to Catholic doctrine. As Aldous Huxley famously noted, Comte's positive religion can be characterized as "Catholicism minus Christianity."[79] The universality desired by the Catholic Church could truly be achieved through a demonstrable, scientific religion.[80] Religion had historically served as a source of order and Comte realized its utility in that regard. However, *progress* could not occur as long as that religion was stuck in the outdated modes of metaphysical and theological philosophizing. Thus, Comte set out to base his doctrine on positive principles that could be accepted by everyone since they would be demonstrable. Faith in an unseen God would no longer be required: the only faith required would be in the science upon which Comte based his system. But since History had already determined its triumph, there was little room for doubt: both positivism and the subsequent Religion of Humanity were destined to prevail. And just as Comte had served as the mouthpiece for the new philosophy, he would serve as the founder of the new religion.

Comte's new religion was to be based on three primary institutions: a doctrine, a moral rule, and a system of worship. All of these would be organized and coordinated by the Positivist church.[81] The doctrine had already been formulated through Comte's own writings, although he never was able to complete important aspects of his project (such as science of morals). The moral rule, or regime, would follow the framework set forth in *Positive Polity*, and the cult would be formed from the initial members of his new church. As Wernick notes, "the Positivist System would provide the scientific-humanist equivalent to what systematic theology had been in the high Middle Ages: it would serve as the intellectually unifying basis of the new industrial order."[82]

The first step in implementing the new Religion would be to disseminate the doctrine. This would be done primarily through a recasting of the educational system; something Comte had initially called for in the *Cours*. The individual's education was to start at mother's knee, "continue in the schools with a revamped curriculum under teacher-priests, and persist in the sermons and ceremonies which Positive Religion would install in a systematic and pervasive ritual round."[83] Those ceremonies included the sacraments of the new religion, which were designed to accompany each stage of life and "through which each servant of Humanity would solemnly rededicate himself to a life of service."[84]

A sweeping reform of education, while helpful, would be insufficient to the task of implementing the new religion. In addition to educational reforms, the major institutions of society had to be reordered. This included the family, the economic system, and the polity, in addition to the obvious reconstitution of the Church:

The family: role-divided, chivalric, extended, replete with children, servants and animals. The sphere of production: cooperative, functionally ordered, justly meritocratic. The polity: reduced to the humanly manageable scale of a small republic, oriented to production not war, and linked to others in an ultimately global confederation. Overarching direction would be provided by a complementary leadership of temporal and spiritual authorities. The former was to consist of bankers, industrialists and engineers from whom, in each republic, a committee of thirty, topped by a triumvirate, would be selected to direct the state. The new "Spirituals," on the other hand, would be the scientists-philosophers-teachers-pastors encadred in the Positivist priesthood itself.[85]

Borrowing from the medieval, Catholic model, the "two leading powers of industrial society were to be not only functionally distinct: each was to have its own form of rule."[86] The priests of the Positivist church were to hold a purely moral, directive power. Unlike the Catholic priests however, the priests of the Positivist church were to care not only for souls, but bodies also.[87] This is because they now held the power of modern science; something the medieval Catholics did not possess.[88] The lay elites, on the other hand, would hold coercive power and coordinate production and distribution.[89] The latter would not overrun their bounds because the church would hold the support of the proletariats and women.[90]

The third institution of the new religion was the cult: "the organized yet effusive worship of Humanity. Under the guidance of the new priesthood, this was to be conducted through public festivals, through worship at the family hearth and ancestral tomb, and through thrice-daily private devotions."[91] Such festivals would follow the Positivist Calendar provided by Comte, which was modeled after Humanity's greatest achievements.[92] As Wernick notes, "If the doctrine was designed to synthesize the understanding, and the regime to synergize action, the cult was to mobilize and canalize that benevolent harmonization of the instincts Comte called *sympathie,* as the proper inspiration of the other two."[93]

The Religion of Humanity was intended as a "subjective" complement to the "objective" foundation of positive philosophy that had been established in the *Cours.* This required a shift in emphasis from the intellectual to the emotional. *Sentiment* would be the driving force of the new religion. As Wernick points out:

> Positivism took sentiments, especially those of the most elevated forms of love, to be the essence of religion. In Comte's general formula: feeling guides action in line with practical knowledge supplied by the intellect. The worship of Humanity was to fix in its adherents a lively impression of such harmonious coordination of the whole human being. The effusions of its rituals would also strengthen the altruistic impulses seen as vital for the correct orientation of thinking and action.[94]

It is important to note that the shift from an objective to subjective synthesis does not represent a fundamental change in Comte's project. His end goal remained the same: to reorganize society on positive principles. The *Cours* was never intended to be the final word on the subject, something that Comte made clear both at the time of its composition and in his later works. Yet, it is important not to ignore the impetus that caused Comte's later works to take on the character that they did. He knew all along that a purely intellectual approach would not be sufficient and that people would have to *believe* in the project, but it was unclear just how he would accomplish that in his early works. The period between 1844–1846 appears to have been when Comte finally formulated the character of the subjective synthesis. His "brief, passionate 'but morally pure' affair with the ineligible Clotilde de Vaux" led Comte to value sentiment more than ever before and seems to have provided the motivation necessary to found his new religion.[95] Immediately following her untimely death, Comte proclaimed himself Pope of Humanity and began systematically constructing the religion.[96] It is to Clotilde that Comte dedicates his later works, and she is given a primary role in the new religion.[97]

The specifics of the Religion of Humanity are outlined in the *Catechism of Positive Religion*. While it is unnecessary to delve into every detail, it is important to note some of the main features of the work. The first point worth noting is the form of the work. Comte presents us with a dialogue, a form he considers to be especially suited for religious instruction:

> The dialogue, the proper form of all real communication, is reserved for the setting forth of such conceptions as are at once important enough and ripe enough to demand it. This is why, in all times, religious instruction is given in the form of conversation and not of simple statement. Far from betraying a negligence excusable only in cases of secondary importance, this form, rightly managed, is, on the contrary, the only mode of exposition which is really didactic: it suits every intelligence.[98]

Comte chose Clotilde as the interlocutress, with an unnamed Priest (presumably Comte himself) of the new religion providing the answers. The choice of Clotilde represents another acknowledgment of her influence on Comte, but more importantly, it represents the elevated role of women in the new religion. While the objective synthesis of the *Cours* was designed to appeal to the intellect of the proletariats, the subjective synthesis appealed to the feelings, or sentiments, of women.[99] Women were to hold a central place in the reorganization of society by providing the initial moral education of its citizens. The education at "mother's knee" was crucial to the proper development of the moral sense required by the new religion. Comte saw his "elevation" of women as a crucial step in the restructuring of society:

> Man ought to maintain woman, in order that she may be able to discharge her holy function. This Catechism

will, I hope, make sensible the intimate connection of such a condition with the whole of the great renovation, not merely moral, but also mental, and even material. Influenced by the holy reaction of this revolution in the position of women, the revolution of the proletariat will by itself clear itself of the subversive tendencies which as yet neutralize it. Women's object being everywhere to secure the legitimate supremacy of moral force, she visits with especial reprobation all collective violence.[100]

In addition to the primary function of providing moral guidance, the social influence of this elevation will "facilitate the advent to political power of the industrial patriciate and of the Positive priesthood, by leading both to dissociate themselves once and for all from the heterogeneous and ephemeral classes which directed the transition in its negative phase."[101] This will culminate in the *peaceful* termination of the revolution in the West, under the direction of the "true servants of Humanity."[102] The retrograde and anarchical, or revolutionary, parties will be eradicated as "all persistence in the theological or metaphysical state" will be "treated as a weakness of brain incapacitating for government."[103]

While the role of women is primarily domestic, the priest functions to cultivate civic morality. Comte bases his whole system on the "concurrence of feeling with reason to regulate activity."[104] The dialogue between Clotilde and the priest represents this union between the heart and the intellect, between the emotional and rational. As Comte notes:

The woman and priest are, in fact, the two indispensable elements of the true moderating power, which is at once domestic and civic. In organizing this holy coalition in the interests of society, each constituent proceeds here in conformity with its true nature: the heart states the questions, the intellect answers them. Thus the very form of this Catechism points at once to the great central idea of Positivism: man thinking under the inspiration of woman, to bring synthesis into constant harmony with sympathy in order to regularize synergy.[105]

The emphasis on unity is once again apparent as Comte looks to reconcile the private and public, as well as the intellectual and emotional, within his system. And to Comte religion, rightly constructed, can provide the ultimate unifying force within society.

While religion provides the means to unification, the structure and character of the previous religions proved inadequate to truly create a universal order. As Comte points out, every religious system "necessarily rests on some explanation or other of the world and of man, the twofold object at all times of our thoughts, whether speculative or practical."[106] What separates the Positive religion from its predecessors is that it can answer such questions adequately:

> The Positive faith sets forth directly the real *laws* of the different phenomena observable, whether internal or external; i.e., their unvarying relations of succession and resemblance, which enables us to foresee some as a consequence of others. It puts aside, as absolutely beyond our reach and essentially idle, all inquiry into *causes* properly so-called, first or final, of any events whatever. In its theoretical conceptions it always explains the *how*, never the *why*. But when it is pointing out the means of guiding our activity, it on the contrary makes consideration of the end constantly paramount; as the practical result is then certainly due to an intelligent will.[107]

This is a remarkable passage in that Comte boldly eschews *ultimate* questions from religion. In the *Cours*, Comte had rejected such questions as falling outside the appropriate parameters of positive philosophy. This was primarily because they produced little, or no, practical results and did not fit into the methodology of positive science. However, the realm of religion would seem to be precisely where such questions would be of paramount importance. While we may not have any "use" for knowing the cause of gravity, we can certainly benefit from knowing how it operates. But in religion, it does indeed matter *why* things work the way they do. Human self-understanding is undoubtedly impacted by such questions, as is human action. If man is created *Imago Dei*, by an omniscient and benevolent God, he is likely to approach society in a different manner than if he is simply the result of a mechanical act of nature. But Comte does not permit such questions because they reflect a "retrograde" mode of thought and tend to lead to discord.[108]

The shift then to a purely immanent religion is what separates Comte's Religion of Humanity from its predecessors. The classical and medieval emphasis on the cultivation of the soul is set aside in favor a materialistic view that places primacy on physical health. This is evident in Comte's insistence on the incorporation of medicine into the priesthood. Without an eternal soul to worry about, or an afterlife to ponder, the emphasis falls on the only existence science assures of: our bodily existence. This elevates politics to first-rate importance and the paradise that used to be reserved for the afterlife must be implemented on earth, if it is to be realized at all. To get a better understanding of this shift, I must briefly outline the features of the new deity of Comte's religion, *Humanity*, as well as the subsequent actions that are intended to honor and strengthen it.[109]

Comte defines the new deity, *Humanity*, as:

> The whole of human beings, past, present, and future. The word *whole* points out clearly that you must not take in all men, but those only who are really assimilable, in virtue of a real co-operation towards the common existence. Though all are necessarily born children of Humanity, all do not become her servants, and many remain in the parasitic state which was only excusable during their education.[110]

This understanding of Humanity has several practical consequences worth noting. The first is that although everyone "belongs" to

Humanity in the proper sense, some will be counterproductive to its flourishing. This group surely includes the retrogrades and revolutionaries, as well as anyone else who persists in asking theological or metaphysical questions. The corollary to this is that a select group of people will truly serve Humanity and it is upon this group that Comte wishes to bestow the greatest honors. The historical nature of Humanity also means that with the passing of time, the present generation will contribute less to Humanity than it will receive from it. This indebtedness to the past leads to "a tendency to strengthen the power of the dead over the living in every actual operation."[111] In place of the classical and Christian distinction between the soul and body, Comte designates an objective and subjective aspect to our lives. The objective simply refers to our bodily existence, the "temporary, but conscious" span of time that we are on earth. The subjective is the "noble immortality, necessarily disconnected with the body" and refers to the "sum of our intellectual and moral functions."[112] Comte even uses the term *soul* to refer to the subjective, but makes it clear that there is no corresponding entity involved. The immortality of the subjective is due to commemoration of the dead by the living.[113] Those who have contributed to Humanity thus live on forever in the minds of the subsequent generations that benefit from their achievements.

Comte insures that the true servants of Humanity will be justly rewarded in the new religion. The Positive Calendar, composed of 13 months (28 days each), is the most prominent example of this emphasis. Each month is named after a great servant of Humanity, and follows a certain theme, and the days within each month are named after someone who likewise contributed to that field.[114] So for instance, the first month is named after Moses with the theme of "First Theocracy" and includes days named after Solomon, Cyrus, Buddha, Romulus, and Muhammad. The calendar serves as a daily reminder of the contributions made by our ancestors and reinforces the notion that great acts will be rewarded well after the termination of our "objective" existence. Public and private worship are intended to reinforce the importance of the "subjective" existence and to encourage due submission to Humanity, or the Great Being.

In order to become a "true organ of the Goddess" of Humanity, the "worthy servant" must pass through nine successive stages, or social sacraments.[115] These include presentation, initiation, admission, destination, marriage, maturity, retirement, transformation, and incorporation. *Presentation* simply consists of a promise by the parents to prepare their child properly to serve Humanity. Drawing from Catholicism, it also requires the selection of an "artificial" couple to offer further spiritual protection of the child.[116] *Initiation* represents the transition to public life as the child passes from his mother's education to the systematic education of the priesthood, at age fourteen. After seven years of instruction under the priest, the young disciple then receives the sacrament of *admission*.[117] At this point, the person is dedicated to the service of Humanity, but has not chosen a specific profession. By the age of twenty-eight, it is expected that a specific path would have been chosen and at that point, *destination* is received. Comte emphasizes that *all* useful professions, whether public or private, are considered viable options.[118] It should be noted that this timeline is intended for males. Females do receive some of the same sacraments, but "their vocation being always known and happily uniform," destination coincides with admission.[119] Thus, the fifth sacrament of *marriage* can be taken by women at age 21, while men must be at least 28. By the age of 42, *maturity* should have been reached and this is when the "second objective life" begins. This is the stage "on which alone depends his subjective immortality. Till then, our life, mainly preparatory, had naturally given rise to mistakes at times of a serious character, but never beyond reparation. Henceforth, on the contrary, the faults we commit can hardly ever fully repair, whether in reference to others or to ourselves."[120] After twenty-one years of maturity, the seventh sacrament of *retirement* follows. At that point, the man is to name his successor, subject to the sanction of his superior, in order to maintain continuity.[121] The eighth sacrament, *transformation*, marks the end of the objective life, or bodily death. Finally, seven years after death, the priesthood determines if subjective *incorporation* will take place. Incorporation is reserved for those who have truly served Humanity and if initiated, the person is buried in a sacred place with "due pomp."[122] Those who are denied incorporation are buried in a "waste place" amongst those who have died "by the hand of justice, by their own hand, or in duel."[123]

The sacraments reflect Comte's systematic emphasis on unity and continuity and further strengthen his insistence on immanent fulfillment. The "immortality" offered by Comte is strictly contingent on the commemoration of the dead by the living so it is crucial that the public and private ceremonies proceed as Comte suggests. It is important not to lose sight of the ultimate end of Comte's religion: social order. The individual holds worth only as an instrument of Humanity, and private worship is designed to inculcate public virtue. Those who do not accept the new religion will necessarily be seen as enemies to Humanity and to progress: there is no room for religious toleration within Comte's system. As Comte notes, "true liberty is essentially the result of due submission."[124] Anyone who fails to submit to the dictates of the new religion is resisting the movement of History itself. Comte's rigidness in this regard led Voegelin to call him a "spiritual dictator of mankind."[125]

Conclusion

Now that I have examined the two main currents of Comte's work, it is instructive to point to a few recurring themes within his system and draw out the political consequences. The foundation of Comte's whole system is rooted in his philosophy of history. Comte's assumptions about the movement of history, and more specifically the law of three states, led him to create an all-encompassing system. The disorder within his society made the creation of *order* a primary concern, yet the character of that order mattered. His understanding of history precluded him from turning to the "retrograde" solution for order; a return to pre-Revolutionary principles was simply not an option. And while religion had served as a source of order, Comte found the major religions, especially Christianity, lacking in several areas. They were based on an outdated, retrograde mode of philosophizing and failed to create a universal consensus. Thus, Comte's challenge was to create a new order; and not just any order, but one that followed the dictates of History and progress.

To bring about such an order, Comte had to restructure the major institutions of society. The first step was to establish the superiority of positivism and reorganize the educational system on those principles. Comte again used his theory of history to establish the justification for such a move, but he also appealed to *utility*. The positive method of philosophizing had led to great advances within the natural sciences and Comte used that to justify its expansion to the social. Comte's whole system is presented as *scientific*, and the success of Newtonian science had given an especially strong, authoritative position to science within society. Nobody could deny the material, tangible benefits that had resulted from the new science. In this sense, Comte had a distinct advantage over Bacon. Bacon's call for a society based on scientific principles was a harder sell to a population that had not yet experienced the transformative power of science. But what Bacon did have was a society that generally embraced Christianity. Comte not only had to promote a new mode of philosophizing, he had to found a new spiritual order.

Comte realized that the founding of a new religion required a modification of the approach that had established the superiority of positivism within the intellectual realm. Yet, he made it clear that the religion was based on scientific, positive principles. The new, universal religion was to be founded on "true philosophy," extracted from "real science."[126] The principles of the religion were not based on revelation, but on science itself. Thus, it was *demonstrable* and would necessarily appeal to everyone, with the obvious exception of those still stuck in the theological and metaphysical modes of philosophizing.

Politically, Comte's dogmatic insistence on the acceptance of positivism meant that toleration would have serious limits. Censorship would be necessary to keep the retrograde and anarchical forces from disrupting order within society. Philosophical questions about the premises of the system, or about formal or final causes, would also be prohibited. Moreover, traditional religions could not be allowed since they were necessarily based on theological and metaphysical foundations. The new religion would provide *immanent* fulfillment. There would no longer be a need to appeal to the "next life" or to any transcendent realm of being. Science would provide the instrument necessary to improve the material wealth of society, while the new religion would provide the spiritual foundation necessary for order.

What should we make of Comte's system in light of contemporary politics? Why concern ourselves with an influential, but often ridiculed, thinker from the nineteenth century? The primary reason for Comte's relevance today lies in his scientistic attitude. Since Comte's time, the prestige of science has only continued to grow, and for good reason. Unfortunately, the rightly earned respect of legitimate science also has been conferred upon politically motivated, pseudo-scientific endeavors. Taking a cue from Comte, politicians have learned to use science as a powerful rhetorical tool. If one wishes to advance a particular agenda, an appeal to science can be an especially effective strategy. In the next chapter, I will examine concrete cases of such appeals in the form of eugenics and Nazi race theory. Once science is viewed as being on the "side" of a particular issue, there is little room left for debate. Comte was certainly not original in his advocacy of scientific ideas, but he was the first to endow it with a religious character. Science was no longer meant to be a mere tool or instrument, but instead it was to be *the* ultimate authority within society. Positivism was not merely an alternative mode of philosophizing; it represented a necessary progression dictated by History itself. And as such, it could not be questioned. Once philosophical questioning has been quelled, along with such inconveniences as liberty of conscience and popular sovereignty, the path to tyranny is paved.

Chapter 5
The Evolution of Scientism:
Marx, Darwin, and Eugenics

"The philosophers have only interpreted the world in various ways; the point however is to change it." –Karl Marx

"I well remember my conviction that there is more in man that the mere breath of his body. But now the grandest scenes would not cause any such convictions and feelings to rise in my mind." –Charles Darwin

Before moving to the constructive part of the study, it is necessary to briefly examine the impact of two contemporaneous figures in the history of scientism: Karl Marx (1818–1883) and Charles Darwin (1809–1882). Marx's political influence is well known, but the scientistic character of his work is often overlooked. While the continuity between Bacon and Comte is apparent, the jump to Marx requires a little more explanation. That an explanation is needed is primarily a result of Marx's explicit rejection of "social utopians" such as Auguste Comte and Saint-Simon.[1] Yet, the critique leveled against those theorists is motivated primarily by pragmatic concerns regarding the means required to achieve their respective ends. Marx simply rejects the idea that a revolution of the mind can transform society.[2] Marx's new men will come *after* the revolution.[3] Unlike Bacon and Comte, he pays little attention to the educational system or to the state of philosophy *per se*. He is concerned with *action*, and violent revolution, not a new mode of philosophizing, is the means by which society will be transformed.[4] For the purposes of this study, it is unnecessary to delve into the intricacies of Marx's program as a whole. The underlying motivation behind Marx's work and the foundation upon which it rests is what deserves close attention. In that regard, Marx's scientific socialism falls much closer to the likes of Comte and Saint-Simon than he was willing to admit.

It is important to recall that the main features of scientism include the dogmatic faith in the methods of the natural sciences, a materialistic worldview, rejection of the *bios theoretikos*, the prohibition of philosophical questions, and an emphasis on immanent fulfillment through the power of science. Marx's system includes all of these aspects, as did those of his scientistic predecessors. What makes Marx's system unique is the political efficacy of his work. Bacon and Comte were both influential in the realm of ideas (i.e., philosophy), but Marx's program had a political influence that is perhaps unrivaled within modernity. It is worth inquiring why Marx's program was embraced on such a large scale, while Bacon and Comte were essentially left with a handful of disciples. According to Voegelin, Marx's success lies not in the inherent superiority of his ideas, but rests in the state of his audience: "Only because the idea was the manifestation of a profound spiritual disease, only because it carried the disease to a new extreme, could it fascinate the masses of a diseased society."[5] In other words, the ideas proffered by Marx's scientistic predecessors had finally permeated society to the point of popular acceptance. Marx did not have to focus on a re-education of society because such a process, to a large extent, had already occurred.

Dialectical Materialism

Marx, like Comte, grounds his system in the dual authority of science and history. His dialectical materialism reflects both of these aspects. Marx and Engels famously claim to have turned Hegel upside down, in order to put dialectics on its feet.[6] Put simply, Hegel's assertion that reality consists in the unfolding of the Idea is rejected outright by Marx who claims, "[T]he ideal is nothing but the material transformed and translated in the head of man."[7] In other words, material conditions serve as the foundation for Marx's system and anything else is simply an abstraction from reality.[8]

According to Voegelin, Marx is not merely rejecting Hegel, but philosophy as a whole. He reaches this conclusion because of Marx's failure to provide a critical foundation for his theory of reality. Marx's "refutation" of Hegel consists in the fact that it does not conform to his own vision of political reality, but that vision is not justified through philosophical reasoning, it is simply assumed.[9] How can this "unphilosophical" position be justified under the Marxian framework? Engels provides the answer through his appeal to the new, materialistic science:

> Modern materialism recognizes history as the evolutionary process of mankind and tries to discover the laws of its movement . . . and also recognizes nature as process and evolution under discoverable laws. With regard to history as well as nature, modern materialism is essentially dialectic and no longer needs a philosophy above the other sciences. This is for Engels the decisive point: when science is occupied with the discovery of the laws of process and evolution, philosophy becomes superfluous. . . . All that remains of philosophy as we know it is the science of thinking and its laws—that is formal logic and dialectics. Everything else is dissolved in the positive science of nature and history.[10]

It is worth noting the distinction between what Engels refers to as "modern materialism" and mechanistic materialism.[11] The primary difference between the two lies in the relationship between matter and motion. Mechanistic materialism separates the two, while dialectical materialism regards them as inseparable.[12] As Engels notes:

> Motion is the mode of existence of matter. Never anywhere has there been matter without motion, nor can there be. . . . Matter without motion is just as unthinkable as motion without matter. Motion is therefore as uncreatable and indestructible as matter itself.[13]

As Hallowell explains, "ultimate reality is matter in motion, a process. Moreover, this is a dialectical process, the reconciliation of opposing movements in an endless effort to achieve a more perfect harmony."[14] The world is still governed by inexorable laws, as in the mechanistic view, but there is a constant evolution involved as matter is seen as "self-moving" and "self-determining."[15] The emphasis is placed more on *process* than *structure*.

In spite of Marx's rejection of the Hegelian dialectic, he accepts Hegel's progressive view of history. But that progress will not take place in the unfolding of the Idea; instead, it takes place within empirical reality. The ultimate *meaning* of the Idea is transferred into intramundane reality, giving ordinary politics an *absolute* character. The *logos* of the Marxian system is the "knowledge of the laws of nature and the possibility, based on such knowledge, of 'letting them operate according to plan for definite aims.'"[16] On this account, "freedom of the will means nothing but the ability of making decisions based on expert knowledge."[17] But how are the ends to be determined? According to Voegelin, the problem of the ends is solved by a theory of the convergence of freedom and necessity:

> The freer the judgment of a man is with regard to a certain question, the greater will be the necessity, which determines the content of the judgment. Insecurity of decision has its source in lack of knowledge; freedom of choice is truly unfreedom because in such indecision man is dominated by the object, which he ought to dominate in turn. *Freedom, thus, consists in the domination of man over himself and external nature that is based on his knowledge of necessity.* . . . The freedom of man advances with technological discoveries. . . . The incarnation of the *logos* has become the advancement of pragmatic knowledge to the point where it has absorbed into its system, and dissolved, the mystery of human existence.[18]

The consequences of this approach are wide-ranging. The tendency to diminish the *bios theoretikos*, as evident from Bacon through the Encyclopedists to Comte, is taken to its logical conclusion by Marx and Engels. "The life of the spirit and the *bios theoretikos* are not merely pushed into the background, they are definitely eliminated. Man will be free when he has achieved perfect knowledge of the external world and with perfect knowledge the problem of purpose, which causes indecision, will have disappeared. . . . The destruction of the substance of man becomes the declared program as a last consequence of the scandal of the *Encyclopedists*."[19]

The transformation that occurs is not one of degree: it is a qualitative change that will extend beyond society and into human nature itself. The "problem of the purpose," along with questions of man's ultimate place in the cosmos will no longer be permitted, as they will have become superfluous. Marx makes this abundantly clear in his *Economic and Philosophic Manuscripts of 1844*. The rise of natural science has allowed for man's true nature to be revealed. It has "invaded and transformed human life all the more practically through the medium of industry, and has prepared human emancipation, however directly and much it had to consummate dehumanization. Industry is the actual, historical relation of nature, and therefore of natural science, to man."[20] It allows us to understand "the human essence of nature or the natural essence of man."[21] Therefore, natural science will become "the basis of human science" just as it has become "the basis of actual human life, albeit in estranged form."[22] Man's "real nature" is revealed by human history and "hence nature as it comes to be through industry is true anthropological nature."[23]

Much like Comte, Marx considers sensory perception to be the basis of all true science. Science must proceed directly from nature and history constitutes a "real part of natural history—of nature's coming to be man. Natural science will in time subsume under itself the science of man, just as the science of man will subsume under itself natural science: there will be one science."[24] Man becomes the immediate object of natural science just as nature becomes the immediate object of the science of man.[25] Marx thus blurs the distinction between man and nature, and reduces the knowledge of both to sensory perceptions and history. Marx's prohibition of questioning follows directly from his understanding of man and nature. Marx responds to an imaginary interlocutor who wants to know "who begot the first man, and nature as a whole?" Marx answers that the question itself is an abstraction: "Ask yourself whether that progression as such exists for a reasonable mind. When you ask about the creation of nature and man, you are abstracting, in so doing, from man and nature."[26] He thus concludes that the interlocutor should give up his abstraction and "you will also give up your question."[27] If the interlocutor responds that he wants to know about the *genesis* (or first cause) of nature itself, Marx replies:

> Since for the socialist man the entire so-called history of the world is nothing but the begetting of man through human labor, nothing but the coming-to-be of nature for man, he has the visible, irrefutable proof of his birth through himself, of his process of coming-to-be. Since the real existence of man and nature has become practical, sensuous and perceptible—since man has become for man as the being of nature, and nature for man as the being of man—the question about an alien being, about a being above nature and man—a question which implies the admission of the inessentiality of nature and of man—has become impossible in practice.[28]

This is a telling passage for it gives us a glimpse into the motivation behind Marx's system. Man cannot be seen as dependent on anything but himself; he is a self-created being. Owing his existence to himself, man no longer needs God because he has become one. For this reason, Marx leaves no role for religion in his system. Marx explains:

> Man makes religion; religion does not make man. Religion, indeed, is the self-consciousness and the self-feeling

of man who either has not yet found himself, or else (having found himself) has lost himself once more. . . . This State, this society, produce religion, produce a perverted world consciousness, because they are a perverted world. . . . The people cannot be really happy until it has been deprived of illusory happiness by the abolition of religion. . . . Thus it is the mission of history, after the other-worldly truth has disappeared to establish the truth of this world.[29]

Along with private property and the family, religion will come to an end after the revolution. On this point, Marx can be accused of naivety since these institutions have been prevalent throughout all of human history and seem to reflect fundamental aspects of human nature. Yet Marx maintains "true" human nature will only emerge once the revolution has occurred. Man has been estranged and alienated not only from his labor, but also from himself. To look at man in his pre-revolutionary state is to look at a different being *in kind* than the one that will emerge after the revolution.[30] Religion is a symptom of the disease that will be cured through revolutionary action. Once deprived of its authority, man will be able to focus on what truly counts: happiness in this world. Thus, just as was the case with Comte and Bacon, politics assumes an *absolute* character.

Revolutionary Action

As previously noted, Marx's method of carrying out his vision varies greatly from the programs of Bacon and Comte. Yet the essential distinction upon which it is based, between thought and action, is embraced by both thinkers. Bacon emphasizes the importance of experiments of light, but reserves his highest praise for experiments that produce fruit or practical utility. His ideal society is one based on the precepts of the new science, developed through his own experimental method, and the transition to that society is presented as peaceful. The revolution is therefore intellectual. Likewise, Comte eschews the contemplative life in favor of practical utility. He criticizes his medieval and ancient predecessors for not focusing on the questions that matter most and outlines a unifying method that will advance science as a whole and directly benefit society. His positive polity will come into existence through the intellectual revolution of his positive philosophy and the complementary spiritual revolution with the founding of the Religion of Humanity. Although Comte and Bacon both emphasize practical utility over contemplation, Marx radicalizes the distinction to the point of abolishing the latter. Comte and Bacon erred by assuming that philosophy or science could transform society. True change can only result through violent *action*. Marx makes this point clear on several occasions. In the *Communist Manifesto*, he concludes:

> The Communists disdain to conceal their views and aims. They openly declare that their ends can be obtained *only by the forcible overthrow of all existing social conditions*. Let the ruling classes tremble at a Communist revolution. The proletarians have nothing to lose but their chains. They have a world to win. Working men of all countries, unite![31]

In his *Poverty of Philosophy*, he echoed the sentiment asking, "Would it, moreover, be a matter for astonishment if a society based upon the antagonism of classes should lead ultimately to a brutal conflict, to a hand-to-hand struggle as its final denouement?"[32]

Marx's realization that violent revolution is the only means through which his end can be accomplished tends to distort his utopianism. For a view that recognizes the necessity of conflict and warfare is generally regarded as realistic. Marx chides the likes of Saint-Simon and Comte for being utopian precisely because they ignore or downplay the necessity of *action*, and more specifically violent action, to achieve their ends. Likewise, Bacon's *New Atlantis* is considered utopian in part because of the peaceful transition that occurs from the old society to Bensalem. Yet, while the means presented by Marx may not be utopian, the ends cannot be absolved of such a charge. Marx does not present us with as clear of a picture of the realm of freedom, the end state of his system, as he does with the realm of necessity. As Voegelin notes, his focus shifts almost exclusively to "tactics" so it is not surprising that most of the emphasis of Marxian scholarship is placed on the revolutionary action itself.[33] But for the purposes of this study, the *intention* behind such action matters, as do the practical effects.

As noted, Marx foresees a different kind of man emerging after the successful revolution. Once the means of production are no longer in the hands of the bourgeoisie and man is no longer alienated from his labor, he will be able to experience true freedom. Under the dictatorship of the proletariat, the last necessary step before true communism, private property, religion, and the institution of the family will all be abolished. As Engels notes, "the care and education of children becomes a public affair; society looks after all children equally, whether they are legitimate or not."[34] Religion is the opium of the people and provides illusory happiness and therefore must be abolished. Likewise, pre-existing moral systems must be abolished since they reflect "class morality" and a "human morality, which transcends class antagonisms," will take its place.[35] Finally, after the dictatorship of the proletariat has successfully expunged the "bourgeois mentality," true communism will be ushered in. It is at this point that "the government of persons is replaced by the administration of things."[36] The coercive function of government is no longer needed; it is relegated to managerial and administrative functions. As Hallowell notes:

> In the new society, class conflict will disappear since there will be only one class, the proletariat. Here all the contradictions of history will be resolved. Exploitation will end. Everyone's wants will be satisfied. Man becomes for the first time the master of his destiny.[37]

Marx's vision then is one where man finally becomes his own master. By uncovering the true laws of history and *acting* in accordance with necessity, man is able to find true freedom. Communism represents

> the positive transcendence of private property, as human self-estrangement, and therefore as the real appropriation of the human essence by and for man; communism therefore as the complete return of man to himself as a social (i.e. human) being—a return become conscious, and accomplished within the entire wealth of previous development. This communism, as fully-developed naturalism, equals humanism, and as fully-developed humanism equals naturalism; it is the genuine resolution of the conflict between man and nature and between man and man—the true resolution of the strife between existence and essence, between objectification and self-confirmation, between freedom and necessity, between the individual and the species. *Communism is the riddle of history solved, and it knows itself to be this solution.*[38]

The end state of communism is nothing short of paradise on earth. There is no need for a coercive government because there will not be any conflict. With material conditions adequately fulfilled, there will no longer be a need for divisive institutions such as the family, marriage, or religion. Man will be free to realize his true nature and pursue whatever ends he desires.

The problem with such a construction is that it is not practically realizable and to pretend otherwise can lead to horrendous political consequences.[39] Human nature does not magically change just because Marx wants it to. The *way* in which Marx constructs his system is particularly troubling. Violent revolution is to occur prior to any of the positive changes taking place. As Voegelin notes:

> Marx did not, like earlier sectarians, create a "People of God" with changed hearts and then lead the People into a revolution. He wanted the revolution to happen first and then let the "People of God" spring from the

experience of the revolution. While for Marx personally the overthrow of the bourgeoisie was senseless unless the revolution produced the change of heart, the historical proof that the overthrow was not the proper method for producing such a change would only come after the revolution had occurred. The pneumapathological nonsense of the idea could not break on the rock of reality before the damage had been done. In the meantime a tremendous amount of disturbance and destruction could be engineered, animated by the pathos of eschatological heroism and inspired by the vision of a terrestial paradise.[40]

In other words, Marx's system led to less than desirable consequences. The idea that violent revolution could serve as the impetus to bring about a new world led to the unfortunate legitimization of violent action.[41] And such action did not lead to the creation of new men: it led to the destruction of millions of men.[42]

So what are we to make of Marx's system in light of contemporary politics and why should we continue to concern ourselves with a movement that has seemingly lost its mass appeal? Marxism, as a viable political ideology, has certainly been on the decline over the past few decades, but that is due primarily to the repeated failures to successfully implement it.[43] The underlying assumptions have not met the same fate. The *scientistic* aspects on which the system is based have continued to permeate the political sphere.[44] The prestige of science has only increased since Marx's time, giving even more credence to policies that are perceived as being scientifically based. In fact, a contemporaneous revolution in science, led by Charles Darwin, had opened the door for an unparalleled merger between science and policy.

Darwinian Evolution

While countless innovations have been made within science over the past century and a half, it is hard to argue that anyone's work has surpassed the social influence of Charles Darwin (1809–1882). Darwin's evolutionary theory, while not entirely original, had far reaching effects within science and society that have continued to this day.[45] Much as Isaac Newton's *Principia* helped shape a new worldview, so too did Darwin's work fundamentally change the way humans viewed the world and their place within it. Because of this fundamental shift, Darwin's theory found import into an array of subjects including eugenics, religion, criminal justice, psychology, economics, and politics. Volumes could be filled on Darwin's influence in both academics and society, but the fundamental presuppositions underlying his theory, along with the political implications of it, are of primary concern. What Darwin's theory purports to explain is almost of secondary importance to what it cannot, or fails to, explain. I will demonstrate how the reductionism inherent in Darwin's theory can lead to deleterious political consequences and also show how it is both the product of, and torchbearer for, scientism.

Phenomenalism and the Loss of Substance

From the onset, it is important to note that in spite of the title of his most famous work, *On the Origin of the Species,* Darwin fails to explain the substantive origin of the species, or of any living being for that matter. Darwin demonstrates how living forms have evolved over time without ever tackling the question of the *substance* of those beings or the ultimate origin of their existence.[46] As Voegelin notes, the answers furnished by evolutionary theory failed to account for "the two fundamental ontological questions of Leibniz: Why is there something, not nothing? And why is the something as it is."[47] Voegelin continues:

> By the time of Kant the problem of evolution was reduced to its phenomenal proportions. And now, in the nineteenth century as if nothing had happened, a new phenomenal theory of evolution, operating with the conceptions of the struggle for life, the survival of the fittest,

natural selection, etc., had a popular success and became a mass creed for the semi-educated. A theory, that assuming that it was empirically tenable, could at best furnish an insight into the mechanics of evolution without touching its substance was accepted as a revelation concerning the nature of life and as compelling a reorientation of our views concerning the nature of man and his position in the cosmos.[48]

The problem then is not theoretical in the sense of a logical flaw in the system or faulty empirics. It is with the fundamental misunderstanding of the relationship between phenomenal relations and substance. In the case of Darwin's theory, it represents the reduction of substance to phenomenal knowledge. Voegelin labels this type of attitude as *phenomenalism*. The preoccupation with the phenomenal aspects of the world, stressed in the natural sciences, has led to the "atrophy of awareness of the substantiality of man and the universe."[49] Voegelin continues, "A theory that in itself might contribute to our knowledge of the phenomenal unfolding of a substance is perverted into a philosophy of substance; the causal relationship of phenomena is understood as an explanation on the level of the substance of life."[50] To get a sense of how this occurs, it is necessary to closely explore the tenets of Darwinism and note some concrete examples of its effects.[51]

Materialism and Evolutionary Biology

As previously noted, Darwin's work represented the popularization of ideas that had already been formulated to a large extent. Malthus's *Essay on the Principle of Population* asserted that the "perpetual tendency in the race of man to increase beyond the means of subsistence is one of the general laws of animated nature which we can have no reason to expect to change."[52] The "struggle for existence" then was a natural condition of life, but Darwin did not necessarily view this as a negative feature: "[I]t at once struck me that under these circumstances favourable variations would tend to be preserved and unfavourable ones to be destroyed. The result of this would be the formation of a new species."[53] To this general process, Darwin gave the name "natural selection." Herbert Spencer, a sociologist, had applied the same principle to human society and coined the popular phrase, "survival of the fittest," which later came to be associated with Darwin's work. Spencer concluded that the government should not interfere with that process by helping the poor as "the whole effort of nature is to get rid of such, to clear the world of them, and make room for better."[54]

In addition to the specific influences of Malthus and Spencer, Darwin embraced the widely accepted materialistic outlook of his time. As West notes, under this view, human rationality, morality, and society are the "products of purely materialistic processes." Institutions such as religion and marriage were direct results of the evolutionary process.[55] The primary difference between Darwin's materialistic account and that of Newton, or even Malthus, was the lack of an appeal to design in nature. While the revelatory aspects of religion had certainly taken a hit with the rise of Enlightenment thought, natural theology had remained as an ally and perhaps last line of defense for Christianity. Natural theology took reason, as opposed to revelation, as its basis. As Brooke notes:

> In England especially, confidence had often been placed in arguments for design, comparing intricate organic structures and their marvelous adaptive functions with the work of human artisans, as in the design of magnificent clocks. Such analogies pointed to the wisdom and power of God, the refinement of whose creatures far transcended anything mere mortals could make. . . . This argument for design had often incorporated the latest science and had been reinforced by it. . . . For Robert Boyle the way the Creator had packed life into the merest mite was awe-inspiring. The physical sciences had also testified to divine precision—in the exquisite calculations made by Isaac Newton's God to ensure that the planets had gone into stable orbits.[56]

On Darwin's own account, his theory undermined such a view of the world: "The old argument from design in nature, as given by Paley, which formerly seemed to me so conclusive, fails, now that the law of natural selection has been discovered."[57] As West notes, "Darwin was no longer filled with the 'higher feelings of wonder, admiration and devotion' when looking at nature."[58] Darwin's own account of this transformation in his autobiography is telling: "I well remember my conviction that there is more in man that the mere breath of his body. But now the grandest scenes would not cause any such convictions and feelings to rise in my mind."[59]

Perhaps the greatest insight into Darwin's view of man can be found in *The Descent of Man*.[60] The Christian view of man was obviously out of the question since Darwin rejected the need for a Creator and thought religion to be a mere stage in the evolution of the human mind.[61] And even if such convictions existed within our minds, how could we trust them? "With me the horrid doubt always arises whether the convictions of man's mind which has been developed from the mind of the lower animals, are of any value or at all trustworthy. Would any one trust in the convictions of a monkey's mind, if there are any convictions in such a mind?"[62] Likewise, there is no reason to believe in the existence of a soul for "Brain makes thought" and thought is produced "as soon as the brain is developed . . . no soul superadded."[63] Darwin further confirms this view by demonstrating "there is no fundamental difference between man and the higher mammals in mental faculties."[64] Love, courage, revenge, suspicion, jealousy, pride, magnanimity, imitation, imagination, reason, and even a sense of humor could all be found in other animals.[65] Language could also be explained away through an appeal to materialistic processes: "[L]anguage owes its origin to the imitation and modification, aided by signs and gestures, of various natural sounds, the voices of other animals, and man's own instinctive cries." Such a process may have originated with an "unusually wise ape-like animal" imitating the "growl of a beast of prey, as to indicate to his fellow monkeys the nature of the expected danger. . . . [T]his would have been a first step in the formation of language."[66] Darwin continues by dispelling the notion that consciousness and abstract thought are distinctly human and finally ends with an attack on the claim that "man was aboriginally endowed with the ennobling belief in the existence of an Omnipotent God."[67] Darwin points out, "numerous races have existed and still exist who have no idea of one or more gods and who have no words in their languages to express such an idea."[68] Moreover, he rejects the need for a Creator and posits religion to be a mere stage in the evolution of the human mind.[69] As John West notes:

> Darwin was willing to admit that there was an "almost universal" belief in some sort of spiritual world among primitive peoples, but he placed such beliefs on the level of a dog who barks and growls upon seeing an open parasol moved by the wind. Just as the dog mistakenly believes that some intelligent agent moved the parasol, primitive peoples surmised that unseen beings were behind all manner of natural phenomena.[70]

In addition to man's mental and physical faculties being reduced to biological and materialistic processes, his moral sense was included. Morality arises because it aids in the struggle for survival: "[T]he origin of the moral sense lies in the social instincts, including sympathy; and these instincts no doubt were primarily gained, as in the case of the lower animals, through natural selection."[71] Morality then depends not on a set of eternal, immutable natural laws, but on material conditions. A preferable morality is one that maximizes biological survival. If this is the case, then the question of free will and human responsibility becomes paramount. Darwin's failure to find a *differentia specifica* essentially means that one cannot judge human action differently from that of animals. As Darwin notes, "at the moment of action, man will no doubt be apt to follow the stronger impulse; and though this may occasionally prompt him to the noblest deeds, it will far more commonly lead him to gratify his own desires at the expense of other men."[72] In his unpublished notebooks, Darwin drew out the implications of this view: "[T]he general delusion about free will is obvious. . . . [O]ne deserves no credit for anything . . . nor ought one to blame others" and the punishment of criminals is "solely to deter others—not because they did something blameworthy."[73] As West notes, Darwin did not think such a fatalistic view would harm society because "ordinary people would never be 'fully convinced of its truth' and the enlightened few who did embrace it could be trusted."[74]

Eugenics

In spite of Darwin's cautiousness in fully drawing out the implications of his system in his public works, others were not so reluctant. Perhaps there is one advantage man possesses over other animals: Man understands the law of nature and has the ability to *control* its effects. Through this knowledge, man has been able to "aid" nature through selective breeding of animals, and with great success. If the difference between man and animal is merely one of degree and not of kind, then why should not this principle be applied to human society? And if human action is primarily understood to be biologically determined as opposed to an exertion of free will, then problems such as crime and poverty can conceivably be eradicated through selective breeding. These assumptions, which Darwin certainly left room for, led to the creation of a new field: eugenics. Francis Galton, Darwin's cousin, coined the term in 1883 and the basic assumption was that the "only solution to social problems was to discourage reproduction by those with undesirable traits, while encouraging reproduction by society's worthier elements."[75] As Diane Paul notes, the movement would soon gain a wide and enthusiastic following that cut across party lines. The evidence seemed to point to the fact that those with undesirable traits were "outbreeding" those with better traits. Human intervention had helped to nullify natural selection as those who would have undoubtedly perished if left to fend for themselves (i.e., the poor and sick), were able to survive through charity. Moreover, the failure of social reforms to correct the situation only served to lend credence to the hereditary, deterministic view. Thus, the only solution to social problems was to be found in selective breeding.

Initially, it was thought that "positive" eugenics could solve the problem. This is the approach Galton took as he encouraged the talented to have large families and outbreed those with less desirable traits. But it soon became apparent that such measures would be insufficient to enact any meaningful change. Simply encouraging the talented to produce large families did nothing to stop those with "undesirable traits" from procreating. Thus, the focus turned to "negative" eugenics and by the turn of the twentieth century, "new views of heredity had converged with a heightened sense of danger and changing attitudes towards the state to make active intervention more acceptable."[76] While Darwin himself was "too imbued with Whig distrust of government to propose that it restrict human breeding," the increasing legitimacy of "collectivist-oriented reform" made government intervention more acceptable.[77] Initially, government intervention took the form of segregating "defectives" during their reproductive years, but as Paul points out, this became too costly and it was found to be much more efficient to simply sterilize those with undesirable traits.[78]

The popularity of eugenics was not limited to Britain: "By 1940, sterilization laws had been passed by thirty American states, three Canadian provinces, a Swiss canton, Germany, Estonia, all of the Scandinavian and most of the Eastern European countries, Cuba, Turkey, and Japan."[79] Eugenics was especially prominent in the United States. The movement would not have gained such momentum without the fundamental shift in American political development that occurred in the early twentieth century. President Woodrow Wilson fully embraced Darwinism to the extent of using it as justification for a new interpretation of society: "in our own day, whenever we discuss the structure or development of a thing . . . we consciously or unconsciously follow Mr. Darwin."[80] West continues:

> According to Wilson, the Constitution betrayed the Founders' "Newtonian" view that government was built

on unchanging laws like "the law of gravitation." In truth, however, government "falls, not under the theory of the universe, but under the theory of organic life. It is accountable to Darwin, not Newton. It is modified by its environment, necessitated by its tasks, shaped to its functions by the sheer pressure of life." Hence, "living political constitutions must be Darwinian in structure and in practice. Society is a living organism and must obey the laws of Life. . . . It must develop." Wilson averred that "all that progressives ask or desire is permission—in an era when 'development,' 'evolution,' is the scientific word—to interpret the Constitution according to the Darwinian principle."[81]

West argues that this opened the door to "much greater regulation of business and the economy, eventually paving the way for the New Deal" and that it naturally led to the idea that government "should scientifically plan and regulate even the most intimate questions of family life."[82] Whether this followed directly from Wilson's progressivism may be debatable, but the acceptance of eugenics, with its Darwinian justification, is undeniable. A few concrete examples of its subsequent impact in America help illustrate the point.

In the United States, eugenics began to gain popular approval as early as 1912. Henry Goddard's book, *The Kallikak Family: A Study in Feeble-Mindedness*, "was received with acclaim by the public and by much of the scientific community" and "interest was expressed in turning it into a Broadway play."[83] Goddard's case study was designed to show that the underclass owed its suffering to heredity, not environment. Goddard traced hundreds of relatives and found "conclusive proof" that one hundred and forty-three "were or are feeble minded, while only forty-six have been found normal. The rest are unknown or doubtful."[84] Goddard posited that people as these were especially dangerous to the racial "stock" because "a large proportion of those who are considered feeble-minded in this study are persons who would not be recognized as such by the untrained observer."[85] Thus, for Goddard, social problems had to be solved through heredity, not charity.

Goddard certainly was not alone in his beliefs. Prominent politicians, scientists, doctors, and even religious leaders adopted similar views. At a meeting of the American Public Health Association, a doctor suggested that all American schoolchildren be given a Eugenics inspection: "As soon as the 2 to 3 percent of all children who are hereditarily defective are determined they should be given such a training as will fit them for the part they are likely to play in life. Then they should either be *segregated* or *sterilized*."[86] A professor at John Hopkins University delivered a similar message a few days later in an address for the International Hygiene Congress, noting importance of "providing for the birth of children endowed with good qualities" while "denying as far as possible, the privilege of parenthood to the manifestly unfit."[87] West points out that the influence of eugenics even spread to the churches, noting the announcement by the Episcopal hierarchy in Chicago that "no persons will be married at the cathedral unless they present a certificate of health from a reputable physician to the effect that they are normal physically and mentally and have neither an incurable nor a communicable disease."[88]

Politically, the increased popularity of eugenics was evident in the creation of organizations promoting its implementation. The American Breeder's Association, Eugenics Record Office, and American Eugenics Society all emerged as prominent organizations. What they all had in common was the firm belief in the scientific grounding of their cause. Their legitimacy arose to a large extent from the pronouncements of prominent scientists at the nation's finest institutions. Edward East, a Harvard biologist, lent his support for negative eugenics:

> Nature eliminates the unfit and preserves the fit. . . . Her fool-killing devices were highly efficient in the olden days before civilization came to thwart her. It is man, not Nature, who has caused all the trouble. He has put his whole soul to saving the unfit, and has timidly failed to

do the other half of his duty *by preventing them from perpetuating their traits.*[89]

East was hardly alone in his views. The return to a primitive state of nature was out of the question so man would have to correct the problem through eugenics. And since "man is an organism—an animal" and "all life is conditioned by the same fundamental laws of nature," then the "same methods that man now employs in producing a high quality breed of dogs, or birds, or cattle, or horses, he must apply to himself."[90] Even prominent inventor Alexander Graham Bell was on board: "All recognize the fact that the laws of heredity which apply to animals also apply to man; and that therefore the breeder of animals is fitted to guide public opinion on questions relating to human heredity."[91]

The utopianism inherent in the eugenics movement was nothing short of astounding. West catalogues a number of striking examples of this seemingly unbounded optimism. Galton had promised that eugenics would gradually raise "the present miserably low standard of the human race to one in which the Utopias in the dreamland of philanthropists may become practical possibilities."[92] In America, proponents noted, "The Garden of Eden is not in the past, but in the future."[93] Proper breeding would mean "little suffering, weakness, sickness, crime or vice." And those who truly embraced eugenics would be empowered like never before: "the people who make eugenics part of their religion and are loyal to its truth . . . will have found the fountain of youth." A new type of human would emerge in the process: "It will take but two or three close-linked generations to make human beings far more superior to us than we are to the apes."[94]

The gaining popularity of such positions led to concrete political changes. As mentioned previously, thirty states adopted sterilization laws by 1940 and restrictions were placed on marriage and immigration as well. The power of the movement even made its way to the Supreme Court, where Justice Oliver Wendell Holmes, an avowed Darwinist, fully embraced eugenics. In the 1927 case of *Buck v. Bell*, Holmes wrote the opinion and upheld Virginia's compulsory sterilization statute:

We have seen more than once that the public welfare may call upon the best citizens for their lives. It would be strange if it could not call upon those who already sap the strength of the State for these lesser sacrifices, often not felt to be such by those concerned, to prevent our being swamped by incompetence. It is better for all the world if, instead of waiting to execute degenerate offspring from crime or let them starve for their imbecility, society can prevent those who are manifestly unfit from continuing their kind. . . . Three generations of imbeciles are enough (referring to Carrie Buck, the woman who was forcibly sterilized).[95]

The chilling decision served to legitimate sterilization laws throughout the country and the sterilization rate increased dramatically as a result.[96] Eugenics eventually lost momentum in the 1940s, but the event that would ultimately sway public opinion against it took place not on American soil, but in Nazi Germany.

Nazi Race Theory

The most striking example of the implementation of eugenics took place in Hitler's Germany. The stage had been set for its acceptance well before Hitler's reign however, and so it is important to understand the context from which it developed. As Diane Paul notes:

Nowhere did the *Origin* have a greater initial impact than Germany, where the book appeared in translation within a year of publication in English. Many scientists endorsed Darwin's theory, which was also widely popularized, most effectively the by University of Jena zoologist Ernst Haeckel. Both liberals and Marxists were enthusiastic. Indeed, Karl Marx's friend Wilhelm Liebknecht reported that, following the publication of the *Origin*, he and his comrades "spoke for months of nothing else but Darwin and the revolutionizing power of his scientific conquests." The response in Germany was so enthusiastic that in 1868 Darwin wrote that, "the support which I receive from Germany is my chief ground for hoping that our views will ultimately prevail."[97]

While Darwinism was used to justify everything from the socialism to laissez-faire, by the turn of the century, it had become increasingly reactionary.[98] The emphasis was placed on the "necessity of competitive struggle" and "modern society was now seen as counter-selective; degeneration could be reversed only through the active efforts of the state."[99] The First World War served to reinforce the "view of war as nature's way of ensuring the survival of the fittest." According to an American evolutionist who served as part of a civilian aid group in the headquarters of the German army in France and Belgium, the German officers defended aggressive militarism by an appeal to Darwinism:

> The creed of the *Allmacht* of a natural selection based on violent and competitive struggle is the gospel of the German intellectuals; all else is illusion and anathema. . . . [A]s with the different ant species, struggle—bitter, ruthless struggle—is the rule among the different human groups. This struggle not only must go on, for that is the natural law, but it should go on, so that this natural law may work out in its cruel, inevitable way the salvation of the human species.[100]

With Hitler's rise to power, the eugenics movement in Germany gained unprecedented support. Paul continues:

> The most extensive and brutal eugenic measures were adopted in Germany. The 1933 Law for the Prevention of Genetically Diseased Offspring, passed soon after Hitler's ascent to power, encompassed a wide range of ostensibly heritable conditions, and applied also to the

non-institutionalized; it ultimately affected about 400,000 people (compared with about 60,000 in the United States). But German *Rassenhygiene* involved much more than a massive programme of sterilization. The Nuremberg Laws barred Jewish-German marriages. The *Lebensborn* programme encouraged racially 'pure' German women, both single and married, to bear the children of SS officers. The Aktion T-4 programme and its various sequels "euthanised" (the euphemism for murder by gassing, starvation and lethal injection) up to 200,000 of the country's institutionalized mentally and physically disabled, sometimes with the tacit consent of their families. The penal system was reformed so that many minor offenders were punished with death in order to counter the dysgenic effects of war.[101]

Such measures were unfortunately but a prelude to the systematic extermination of millions of Jews that would follow in subsequent years under Hitler's reign. But we must ask *how* such a system could be accepted in the first place. What allowed eugenics and social Darwinism to rise to such prominence? In short, the answer lies in the "new" view of man, in biological reductionism, which posits man as nothing more than a product of mechanical and biological processes. To understand how this transformation occurred, it is instructive to examine Voegelin's penetrating analysis of the "race idea."

Writing in the shadows of Hitler's Germany, Voegelin had the "benefit" of directly observing the implementation of the various eugenics programs. Astonishingly enough, Voegelin was able to publish two books in Hitler's first year in power (1933, *Race and State, The History of the Race Idea*) in which he heavily criticized the ideas that the Nazis used to justify their programs.[102] Voegelin contended that the race theory of the Nazis was "an image of destruction" and represented the decay of race theory in general.[103] In the introduction to *The History of the Race Idea*, Voegelin laments, "The knowledge of man is out of joint. Current race theory is characterized by uncertainty about what is essential and a decline in the technical ability to grasp it cognitively."[104] Such a deficiency is characterized by the change in the view of man; or in Voegelin's terms, the shift from a Christian primal image to a post-Christian image.[105] Voegelin explains the character of the Christian image:

> The Christian image raises man out of nature; though it presents him as a creature among other creatures, as a finite being among others, it nevertheless juxtaposes him to the rest of nature; he stands between God and the sub-human world. This intermediate status is not determined by a unique formative law that would constitute man as a self-contained existence but by his participation in both the higher and lower world. By virtue of his soul, man is united with the divine *pneuma*; by virtue of his body, his *sarx*, he partakes of the transitoriness; his existence is "inauthentic." The condition of his existence is that of being lost, an existence from which he must be freed in order to ascend to the realm of his true existence with his "authentic" nature.[106]

In other words, an emphasis is placed on man's ability to transcend his bodily existence. Man is characterized by his ability to commune with God and the care of his soul is what ultimately matters, as the body represents nothing more than a transitory hindrance to his true end.

The shift from a Christian to post-Christian image of man is "not marked by a sharp break, a clear end and a new beginning; rather, it is a blending of one image into another, a fading out of one and simultaneous intensifying of the other."[107] According to Voegelin, the shift begins with the "new understanding of living nature" that results from the proliferation of the natural sciences; especially in fields such as zoology and botany.[108] The systematic classification of living nature leads to the "question of the significance of the body and its diverse forms for an understanding of man."[109] Should man's essence be defined in a similar way to animals or as something unique? Can a systematic, "natural" classification of man be reconciled with the Christian view? As Voegelin notes, these are the prominent questions that first arose with the systems of Linnaeus and Buffon and were subsequently explored by the likes of Kant, Blumenbach, and von Herder.[110]

Voegelin argues that the newfound knowledge of nature, along with the questions it raised, led to a new way of viewing man and his place in the world. Voegelin distinguishes two phases in this process:

> In the first phase, the Christian image of nature was dissolved. Plants and animals had been seen as creatures of God, as materials shaped and ensouled by the hand of a master craftsman. The living substance presented itself as the medium in which a plan had been realized; organized matter was understood as a construct, a machine, an instrument embodying the ingenious idea of its builder and moving according to this idea. In the eighteenth century this image of the machine was gradually changed into that of a substance carrying the law of its construction within it, a substance that is not created or animated from outside but that is itself a primary force, a substance that is not given its life from the outside but lives out of the wellsprings of its own aliveness.

To this first stage, Voegelin appropriates the phrase "internalization of the body." The "formative drive" takes precedence as emphasis is placed on the phenomenal aspects of life. The life of the body thus gains primacy over the life of the spirit.

Prior to this stage, the "development of the organic form" was not seen as a primary phenomenon, but as a "machine put together from the outside." Voegelin points to the work of English botanist and zoologist, John Ray as an example of this viewpoint.[111] Ray refused to take a position on the question "of whether all individuals are already present in the first individual, with the act of procreation being merely the impetus to their growth, or whether God intervenes each time at the moment of conception to ensoul the matter that in itself is lifeless," but considered both views, *preformation* and *epigenesis*, to be potentially valid.[112] According to Voegelin, what these views share in common is much more important: they both assert that the "living form is not alive out of the wellspring of its own vitality."[113] The organism is considered as an "artifact" and owes its creation to rationally planned reason.[114]

The shift to the "internalization of the body" is evident in the reinterpretation of mechanism as organism and in "finitization."[115] Whereas at the beginning of the eighteenth century, the concept of the organism was tied firmly to mechanism, by the end of the century, organism had acquired a new meaning. Voegelin demonstrates this gradual change by examining the works of Leibniz, Wolff, and Oken. The transcendental element is still firmly present in Leibniz's work, "God alone stands above all matter since he is its creator," but while the body is still a "species of divine machine," the distinction between the plan and the material in which it is executed is lost.[116] This is due to Leibniz's idea of monads, which were understood as the organizing principle of the body.[117] Voegelin explains the consequences of such a construction:

> The monad is neither body nor soul—that is, neither dead matter nor ensouling principle—but living body, an entity that does not fit into the categories of body and soul but that Leibniz has to comprehend in terms of these categories. That is why he ends up with the curious construction in which the organism must be seen as a composite of material elements under the rule of a soul while at the same time both the material elements and the soul are also monads—that is, neither material nor soul but the desired third element, the unities of life. Thus, a machine in the mechanical sense emerges in these speculations that in its inner structure is not a machine in this sense after all; instead it is an organism.[118]

Wolff, writing in the middle of the eighteenth century, maintains the language of transcendence in his construction by continuing to refer to an immaterial soul, but the emphasis is shifted to the "strong inner animation" of the immanent body. Wolff, while still maintaining a quasi-Cartesian dualism, "questions whether there can be any doubt that the mechanical actions are nothing more than an insignificant appendage of the animal itself."[119] The emphasis on the inner drive of the organism itself, however, represents a bold step towards the internalization of the body.

By the end of the eighteenth century, the shift to the internalization of the body was nearly complete. Voegelin refers to the work of Oken to illustrate this shift. Oken works within the basic framework provided by Leibniz, but replaces the notion of monads with protozoa, or infusoria. Like the monads, "the protozoa are the elements of the organic world of which all higher animals are composed." But the protozoa, unlike Leibniz's monads, "are not considered the simple substances of being as a whole but are specific organic elements." Oken, although struggling to break out of the mechanical language of his time, makes it clear that his formulation is organic, not mechanistic: "The union of protozoa in the flesh should not be pictured like a mechanical adhesion of one tiny animal to another, as in a heap of sand where the grains are not connected in any other way than simply lying next to each other—no! Analogous to the disappearance of hydrogen and oxygen in water, mercury and sulfur in cinnabar, this union is a genuine interpenetration, a growing together, a fusion of all these protozoa."[120] While Oken retains certain aspects of Leibniz's system, the natural unity of the organism has become much more "inward." The organic idea represents a fusion of the two "realities" originally explored by Ray: "the essence as operative cause of the phenomenon and the essence as the essentiality that is discernible for the observer in the traits that form the type."[121] In other words, the organism carries its own design from within and possesses an internal generative principle.

With the shift in the idea of the organism, from transcendent to immanent, came a new understanding of the species and its "constancy through the generations." Formerly, the transcendent view had ascribed a fixed beginning to the creation of the species. The "fixity" of the species could easily be explained under this view as the manifestation of the will of a Creator.[122] When the Christian worldview began to dissolve, the "starting point" of the species was lost. As Voegelin explains:

> And while the similarity of individuals had been understood as caused by a similar pressure of the divine hand, the theory of the fixity of the species was now also shaken. The succession of generations no longer had a finite beginning in the creation or an origin of its specific laws; instead, the succession could be traced from any individual back into infinity without this regression coming up against a point of origin for the law of the species. The result was a peculiar, undecided state. The concept of creation was replaced by the idea of infinity.[123]

According to Voegelin, the idea of infinity was quickly discarded because it rendered the idea of a fixed law, formerly posited at the beginning of creation, nonsensical. The search for a beginning would merely lead to an infinite regress. This dilemma led to the "speculative leap to the lawfulness of the species as a real cause that is at work in all individuals of a species, thus necessarily also in the one currently under observation, without having to be traced back to preceding ones." Voegelin refers to this process as the "finitization of the law of the species." The convenience of such a concept is that it fits well with the new concept of organism that had developed throughout the eighteenth century. The unity of the species and of the individual organisms that compose it is thus explained through an "inner law."

The second phase in the shift from the Christian to post-Christian image of man is characterized by the "internalization of the person," as the "image of man as an earthly, self-contained, unified figure develops."[124] Voegelin points to the work of the physician, Carl Gustav Carus, to show how the new image of man is taken to its logical conclusion. Using Goethe as his model, Carus posits the notion of the "well-born" man. The soul and body are no longer seen as oppositional figures, but as complements to the edification of the whole person. The body is seen not as a transitory hindrance, but as the "foundation without whose good constitution a wide-ranging and free unfolding of the spirit is impossible."[125] Carus's characterization of the Goethean ideal extends beyond the man himself to "the line that begat him:"

> The competent, somewhat pedantic, but thoroughly dis-
> tinguished and honorable nature of the father, the deli-
> cately humorous, genuinely feminine character of the
> mother who was more or less high-spirited and lively
> into old age, have laid a basis here that could well be-
> come the element for allowing an idea of life to express
> itself that at one time could prove itself in many respects
> one of the perfect blossoms of humanity; Goethe was in-
> deed what is said of so many but what so few truly are—
> a well-born man.[126]

Voegelin is careful to point out that Carus, unlike the contemporary race theoreticians, understands the implications of the idea of the well-born man in a way that is "far removed from the barbaric nat-ural scientific dogmatizations of modern eugenics, which narrows the concept to apply only to particular physical conditions."[127] Carus is concerned with both physical and spiritual health and posits that a disease in one can hinder the other. The well-born man is one who can overcome diseases of the spirit and body, through his own inner strength and fortitude.

The political implications of Carus's well-born man are far reaching. The fact that such a man exists, concretely in the person of Goethe, presents the first step in the exploration of the genetic inequality of men. And since Goethe's well-born existence depends significantly on his ancestry, the notion that a particular race could have inherent advantages over others is not far behind. For Carus, race theory could help us understand "how such a powerful individuality as that of our Goethe could emerge only from one tribe, which in itself was already a higher one and which therefore generally already promised its members an outstanding and powerful spiritual development."[128] Carus finds, among the numerous races, a rank ordering of development of physical and spiritual capacities. Once it is accepted that all races are not equal, and that the development of great individuals depends to an appreciable extent on ancestry, the "jump" to eugenics becomes much more intelligible. The assertion of an *übermensch*, of a man that stands above the rest, led to the dream of a super race, with disastrous results for humanity.

Darwinism in the Twenty-First Century

In spite of the apparent failure of the eugenics movement, many of the assumptions on which it was based are still present in the twenty-first century. Although advances have been made in the fields of genetics, biology, and other related life sciences, Darwinism remains prevalent. In fact, some of the debates that Darwin encountered in the nineteenth century are still raging today, especially in America. More specifically, the question of what evolution actually explains has become paramount. The result of the confusion is in large part due to the influence of scientism. Darwin never explained the origin of life itself, although he held out hope that the question could be answered someday.[129] His theory merely explains how life, once given, evolves. It speaks to phenomenal relations and remains silent on questions of substantive origins. But the emphasis on biological determinism, along with the new view of man, leads to the misidentification of parts for the whole.[130] The emphasis on the biological, or the physical, leads to a devaluation of the spiritual. Darwin, while not denying the existence of a spiritual element in man, nevertheless attributes its development to biological processes.[131] Indeed, as noted previously, Darwin goes through a long list of attributes that had been posited as reasons for man's uniqueness and shows that each can be explained by his theory of natural selection. The annihilation of a *differentia specifica*, along with the undermining of natural theology and the need for a Creator, exemplifies the new view of man that Voegelin had outlined in *The History of the Race Idea*.

Darwin's work served to legitimate and popularize the "biological" view of man.[132] As demonstrated in the case of the eugenics movement, that view of man led to devastating political consequences. It would be unfair to place blame on Darwin for the horrendous actions of the Nazis, or the embarrassingly inhumane sterilization laws in America, but it is less of a jump to assert that the view of man proffered by his work was heavily influential in the acceptance of such policies. What is more striking is not necessarily the content of Darwin's work, but the prestige that was ascribed to it. Nearly every political faction appealed to Darwin's work in some way to justify its positions. Everything from laissez-faire to progressivism to Marxism to Leninism sought legitimacy in the works of Darwin. It was not uncommon to see *both* sides of a particular issue bolstered by Darwinian appeals. The reason for those appeals is apparent: science offers legitimization. The rhetorical power of science can hardly be overstated.

Darwinism remains a dominant force today. Advances in genetics, such as the discovery of DNA, have fit rather comfortably into the Darwinian framework. While eugenics has fallen out of favor, more so in terminology than in spirit, the idea of human control over genetics has only increased. This is evident in "gene therapy" and in pre-natal tests, which allow for the diagnostics of harmful traits or diseases. The Human Genome Project will offer "thousands of pre-natal tests within the next decade or so" and presumably such information will allow couples to decide whether to terminate life or continue the pregnancy based on that genetic information.[133] This offers much more control over transmitted traits than mere sterilization, as seen in the "negative" eugenics of the early twentieth century. The problem remains as to *how* these decisions should be made. If biological reductionism and determinism remains prevalent, then a strong case can be made for fully embracing these advances. What is to stop private individuals, or even governments, from intervening and preventing certain traits from being transmitted? Crude measures such as forced sterilization and population extermination may give way to "gentler" procedures, but the underlying principle remains. The field of Bioethics is designed to help address such questions, but can a field borne out of the intellectual climate that led to the creation of eugenics be expected to quell the negative effects of it?

The questions regarding science's role in the twenty-first century are not unlike those raised in the exploration of Bacon's *New Atlantis*. In an age of unprecedented technological advances, the questions have only gained importance. Should all scientific exploration be permitted or are certain innovations to be discouraged? And if so, on what basis can that decision be made? The power offered to man by technology is undeniable, but the question remains as to how that power is used. To the question of who gets to decide, Bacon and Comte would certainly point to the "scientific" statesman. And in the twentieth century, a strong push was made to let "managerial experts" make such decisions.[134] But in light of the events of the twentieth century, we can rightly question the efficacy of such leadership. The solution lies in political science, the discipline designed to provide the practical wisdom necessary to make such decisions. Yet, modern political science has generally modeled itself on the very sources that have contributed to the problem of scientism. It has taken the model of the natural sciences as its own and in doing so, has lost its way. In order to confront the problems of our age, and more specifically scientism, a new science of politics is necessary.[135] Thus, I will now focus my attention on that task and outline what such a science would need to encompass in order to address the concrete political problems of our age.

Chapter 6
Towards a New Science of Man

"So we must examine the conclusions we have reached so far by applying them to the actual facts of life: if they are in harmony with the facts we must accept them, and if they clash we must assume that they are mere words."[1] –Aristotle

Before outlining what a new science of politics should encompass, I will first outline the reasons why such a seemingly radical step is necessary. The central argument of this work is that scientism stands as the key crisis of our age. Yet, as I have shown throughout the analysis, the problem is not a new one. Scientism is firmly rooted in the modern project, as demonstrated in the analysis of one of the founders of modernity, Francis Bacon. The political problem of scientism did not become acute until science itself had gained sufficient prestige within society. Newton's work served as the impetus for this transformation and the French, beginning with the Encyclopedists through Saint-Simon and finally to Comte, fully capitalized on the situation. Riding the wave of momentum created by Newton, they proposed a radical recasting of the educational system. As part of that reconstitution, the study of man was "positivized." Natural science served as the model on which all science, human affairs included, was to be based. This shift led to the assumption that law-like generalizations could be found in the social sciences. It also meant that certain types of questions could no longer be asked, an issue which was discussed at length in the analysis of Comte's *Cours*. The diminishment of the *bios theoretikos*, along with the prohibition of philosophical questioning, led to a new view of man. No longer grounded in a classical or Christian philosophical anthropology, the new view of man found its bearings in natural science. Popularized by Darwin, the "biological" view of man led to deleterious political consequences. Natural science provides the knowledge necessary to modify, and to an extent control, nature. To that end, it has been incredibly successful. One needs look no further than modern medicine and technology to realize our indebtedness to science. However, natural science does not provide us with guidelines on how that power is to be wielded, nor does it aim to do so. The same technology that can provide power for a whole city can also be used to destroy it. What is needed then is a science of human affairs, one that nurtures the practical wisdom necessary to make such decisions. Properly speaking, such issues belong in the domain of political science.[2] Yet, as previously noted, contemporary political science has found itself impotent in the ability to address such questions. First I will demonstrate *why* contemporary political science is not up to the task, and then I shall address the primary question of what its replacement must encompass.[3]

Contemporary political science owes its origin to Enlightenment thought, and more specifically, to the work of Auguste Comte. The overall assumption was that law-like generalizations of social behavior could be discovered, just as they had been in physics by Newton. If mechanical laws could be used to explain the inner workings of nature, it was assumed that they could also be applied to humans. Following Bacon's lead, the Aristotelian framework for the interpretation of human action was abandoned in favor of the model of natural science. No longer would human action be defined in terms of "intentionality" or "purpose."[4] The teleological nature of Aristotle's work fails because it relies on final causes; something that the model of natural science does not concern itself with.[5] Once questions of purpose are excluded, ethics loses its grounding.[6] The very questions that were crucial to the health of the polity under Aristotle's account become meaningless under the new model of social science proffered by the likes of D'Alembert and Comte. This is not to say that the *aims* of those thinkers are not noble. Indeed, the motivation behind the new science is directed towards the creation a peaceful, prosperous society.

The problem is that those aims cannot be met, due to a fundamental misunderstanding of human nature and of the power of the newly created "social physics." In other words, social science cannot do what it aims to do.

The fact that social sciences in general, and more specifically political science, have not been able to produce the law-like generalizations akin to the natural sciences is in itself not a sufficient reason to replace the paradigm.[7] The reasons *why* it has not been successful are what deserve attention. Alasdair MacIntyre's argument on this account is one worth examining. MacIntyre addresses the two most common defenses offered on behalf of social science's progress, or lack thereof: social sciences are "young" and merely "probabilistic." To the first defense, MacIntyre simply appeals to the historical record and shows that social science has been around for hundreds of years and has failed to produce anywhere close to the record that has been produced in any given natural science.[8] To the claim that social science only seeks to produce probabilistic generalizations, and therefore that counterfactuals do not discredit the theory, MacIntyre asserts that the kind of generalizations that are actually produced are nothing like those in the natural sciences. For in the natural sciences they offer probabilistic generalizations, but they do so with "universal quantifiers—quantification is over sets not individuals—they entail well-defined sets of counter-factual conditionals and they are refuted by counter-examples."[9] Without such conditions, there is nothing "law-like" about the generalizations, whether they are posited as probabilistic or not.

It is important to qualify MacIntyre's argument lest it appear that he is simply impugning the whole enterprise and its practitioners. Social science practitioners are not failing to produce the law-like generalizations that they hope to because of any intellectual deficiency or lack of effort. Their task is simply impossible because the assumptions on which social science is based are faulty. Those assumptions can be traced all the way back to Bacon. As noted in the previous section on Bacon and Machiavelli, the primary difference between their philosophies is the way they view *fortuna*, or chance. Machiavelli laments the impact that chance plays in human life and points out how even the most virtuous prince can fail because of it. While Machiavelli does allow for the possibility of humans to minimize its impact, he realizes that it will always serve a crucial role in human affairs. On the other hand, Bacon's new science is designed to do what Machiavelli thought impossible—control chance.

MacIntyre points to Machiavelli as an alternative foundation to social science precisely because Machiavelli correctly diagnoses the role that chance plays in human affairs.[10] Machiavelli knew "no matter how good a stock of generalizations one amassed and no matter how well one reformulated them, the factor of *Fortuna* was ineliminable from human life."[11] This inevitably means that unpredictable counter-examples will arise and it is the frequency of such occurrences, which makes the production of "law-like" generalizations dubious. To further illustrate the point, MacIntyre outlines four systematic sources of unpredictability in human affairs. The first is "radical conceptual innovation." We cannot predict a radical conceptual innovation, such as the discovery of the wheel, because to do so would require an explanation of the concept that only comes about through the innovation.[12] In other words, if one were to predict the discovery of the wheel, he would have already discovered it. The second type of unpredictability relates to the unpredictability of an agent's own actions. As MacIntyre explains, "decisions contemplated but not yet made by me entail unpredictability of me by me in the relevant areas."[13]

The third source of systematic unpredictability is of particular importance for this study since it deals directly with a methodology that has become prevalent in political science: game theory. Game theory represents the most prevalent attempt within the discipline to provide law-like generalizations. MacIntyre characterizes it as follows:

> Take the formal structure of an n-person game, identify the relevant interests of the players in some empirical situation and we shall at the very least be able to predict what alliances and coalitions a fully rational player will

enter into and, at a perhaps Utopian most, the pressures upon and the subsequent behavior of not fully rational players.[14]

MacIntyre outlines three reasons why the aims of such a theory necessarily fall short. The first concerns the "indefinite reflexivity" of game theory situations. This refers to the fact that both sides will have to predict what the other will do, yet each actor will try to misrepresent his own intentions. The fact that the other actor is trying to misrepresent his intentions is known by both sides, and thus a situation arises where neither actor can be certain as to the intentions of the other. While MacIntyre acknowledges that this alone may not prove insurmountable, the second obstacle, imperfect knowledge, makes it highly unlikely. Each actor has an incentive to "misinform" the other side. This proves problematic for any external observer for "success at misinforming other actors is likely to be the successful production of false impressions in external observers too."[15] The final obstacle that prevents game theory from successful predictions is that in the real world, it is rare that simply one game is being played. It is likely that several games are being played; even by members of the same group. In other words, real-life situations are often too complex to "fit" into a formal scheme.

The final source of "systematic unpredictability" is pure contingency. MacIntyre points to Napoleon's cold at Waterloo as an example of this. Had he not fallen ill, command would not have been delegated to Ney, and the Imperial forces may have prevailed.[16] The beginning of World War I can also be attributed to pure contingency, as Archduke Ferdinand's assassination was hardly the result of a well-planned scheme. Nearly every major conflict has contained elements of contingency, and they often play crucial roles in determining the outcome. Taken together, the sources of systematic unpredictability present a daunting task for anyone who wishes to explain, and especially predict, human affairs.[17]

MacIntyre does concede that there are elements within the social world that can be predicted. "Systematic predictability" consists of four primary elements. The first is that we must schedule and coordinate our social actions. People generally go to work, eat, and sleep at similar times.[18] The second source of systematic predictability arises from statistical regularities (suicides peak around Christmas, most catch colds in winter, etc.).[19] It is important to note that we often have no concrete causal explanations of why these occur; further strengthening MacIntyre's point that predictability does not necessarily entail explicability.[20] The third and fourth sources of systematic predictability are the causal regularities in nature (seasons, storms, earthquakes, etc.) and causal regularities in social life.[21] In spite of these factors, the previously mentioned sources of unsystematic unpredictability remain prevalent. Chance is something that cannot be eliminated and unpredictability is a necessary condition of social life.

So where does social science stand if MacIntyre's argument holds? He suggests that social science will never attain the law-like generalizations that it currently seeks. It must restrain itself to offering inductively-based generalizations, not unlike those proffered by Machiavelli. Likewise, there will be no well-defined set of "counterfactual conditionals."[22] Moreover, he suggests that social scientists pay closer attention to error. If it is not randomly distributed, as is often assumed, then we may learn more about the role chance plays in specific parts of the social world.[23] But the more important implication of MacIntyre's critique is not that it "undermines" social science as presently constituted, but that it questions the very social order that has ascended with it. MacIntyre characterizes twentieth century social life as the "concrete and dramatic re-enactment of eighteenth-century philosophy. And the characteristic institutional forms of twentieth-century social life depends upon a belief that some of the central claims of that earlier philosophy have been vindicated."[24] MacIntyre's argument is designed to show that those claims have not been vindicated and can never be since they fail to recognize the inherent complexity and unpredictability of human affairs.[25]

Before turning to the constructive part of the study, it is necessary to examine one more facet of social science that MacIntyre only hints at. The adaptation of natural scientific methodology in the social sciences could not have occurred without a fundamental *ontological* shift. That such a shift occurred is hardly disputable, but the nature of that shift has been obscured. This is due to the fact that ontological questions were subsequently "banned" by those who initiated the shift.[26] More specifically, a new view of man was adopted. And it was on the basis of that view of man and his place in the world that the epistemological assumptions, addressed at length by the likes of MacIntyre, Hayek, Andreski, and Voegelin, of social science were developed.[27] The methods of the natural sciences could only be transferred to the study of man if man was firmly rooted in the "natural" or "material" world. As I have noted throughout this study, that view was adopted in part as a rejection of classical and medieval philosophy, concretely in the work of Bacon and Comte, and also as a result of the astounding success of natural science (especially with Newton's work). The "materialistic metaphysic" allows for a purely phenomenal epistemology.[28] Nature is asserted as the whole of being: to exist is to take part in the physical existence of this world. Man's physical nature is taken to be the whole of his existence and knowledge is attained through sense perception. Once these ontological and epistemological assumptions are adopted, it is of little surprise that the methods of the natural sciences are implemented. If man is simply another part of the natural world, then there is no reason he cannot be studied in the same manner that other living beings are examined. My central claim however is that such an approach does not and cannot work. And this is due not only to the epistemological questions raised by critics such as Voegelin, Kolakowski, Andreski, and MacIntyre, but more importantly, because it bases its claims on a faulty ontology. Thus, even if it succeeds in its epistemological aims (which I maintain that it does not), it wrongly identifies the part(s) as the whole. And this mistake is what ultimately renders social science, and contemporary political science, unable to fully address the major political problems of our age. Now that some of the critical issues that have plagued political science, and social science in general, since its inception have been identified, I can now turn to the constructive part of the study and suggest some corrective measures.[29]

From the onset, it is important to acknowledge an inherent difficulty in proposing such a solution: language has been co-opted by the very system that I propose to replace. More specifically, terms such as reason, experience, facts, empiricism, and perhaps most importantly, science, have taken on different meanings within modernity. Bacon and Comte both appeal to experience and reason, but they have something different in mind than did Plato and Aristotle. Thus, a crucial part of the task will be to clarify the different meanings of these terms and to ultimately recover the elements that were lost with the transition to modernity. It is necessary to begin with perhaps the most important of these terms, *reason*, for all science (whether of the ancient or modern variety) depends on it for its justification and implementation.

Reason

The one thing that nearly every philosopher and scientist will agree upon is that reason is crucial to his enterprise. Yet, we would find less agreement if we asked them to define reason. This is in part due to the fundamental shift that occurred with the inception of modernity whereby reason acquired a much more "practical" bent.[30] The *bios theoretikos*, or contemplative life, was diminished in favor a practical utility. Reason, instead of being considered the pinnacle of the soul as in Plato and Aristotle, became a mere means to an end within modernity. And instead of ordering the soul, reason became the slave of the passions and was relegated to a tool used in their service.[31] By the time of Darwin, reason had become nothing more than an evolutionary tool to aid in survival and serve the self-interest of human beings. And unlike Aristotle, who posited reason as the *differentia specifica*, Darwin was quick to ascribe it to other living beings. Reason could help us, and some other living beings, secure certain ends, but it would have little, if anything, to say about which ends to pursue.

There are at least two reasons for this change in the meaning of reason. The first is the tremendous success of the natural sciences in providing tangible benefits. Medicine and technology provided "proof" of the efficacy of the new conception of science and its implementation of reason. Secondly, the gradual diminishment of Christianity's influence led to an emphasis of the physical over the spiritual, which subsequently meant that the body gained primacy over the soul.[32] Aristotle had praised the *bios theoretikos* as the happiest life because it requires the utilization of man's most divine capacity.[33] To the extent that we embrace that capacity and engage in contemplation, we are "immortalizing." Like Plato, he places reason at the pinnacle of the soul. And to care for the soul is far more important than the body as Plato makes clear throughout his works.[34] Once the soul's existence is denied or placed in the service of the body, reason necessarily must be redefined. And in this redefinition, reason loses any connection to the divine and ceases to be the constitutive part of man's nature.[35] It is my argument that this redefinition of reason was not justified, and it has led to *stasis*, or rigidification, within the political order. To understand why this is the case, it is necessary to examine reason in the full, classical sense.

Eric Voegelin offers a penetrating analysis of reason's role in antiquity in his essay, *Reason: The Classic Experience*. Voegelin posits reason as an epochal event in the history of mankind. It represents a "process in reality in which concrete human beings, 'the lovers of wisdom,' the philosophers as they styled themselves, were engaged in an act of resistance against the personal and social disorder of their age."[36] It serves as an ordering structure in the *psyche*, or soul, of man and is borne out of a concrete resistance to disorder, both within the individual and in society. To this particular kind of reason, Plato and Aristotle gave the name *nous*. As Voegelin notes, *nous* serves as the "cognitively luminous force that inspired philosophers to resist and, at the same time, enabled them to recognize the phenomena of disorder in the light of a humanity ordered by *nous*."[37] Therefore, noetic reason constitutes both the "force and criterion of order."[38] In other words, it provides us with the cognitive tools necessary to recognize disorder while also providing us with the model of proper order to correct it.

Voegelin is careful to point out that the discovery of *nous*, although epochal in the philosophical sense, did not magically cure the political ills of the time:

> Reason in the noetic sense, it should be understood, does not put an apocalyptic end to history either now or in a progressivist future. It rather pervades the history, which it constitutes with a new luminosity of existential order in resistance to disordering passion. Its *modus operandi* is not revolution, action, or compulsion, but persuasion, the *peitho* that is central to Plato's existence as a philosopher. It does not abolish the passions but makes reason articulate, so that noetic consciousness becomes a persuasive force of order through the stark light it lets fall on the phenomena of personal and social disorder. To have raised the tension of order and disorder in existence to the luminosity of noetic dialogue and discourse is the epochal feat of the classic philosophers.[39]

The reason that *nous* cannot provide a magical political cure is because of the fact that each individual must first overcome disorder in his own soul.[40] The tension of existence, to use Voegelin's terminology, requires "balance of consciousness."[41] The "aperionic depth," or nothingness, holds an equally prominent place in man's nature, as does the divine *nous*. Furthermore, man is not a "disembodied psyche ordered by reason."[42] He possesses a corporeal nature and experiences the pull of the passions. Thus, while criticism of the radically immanent conception of man proffered by the likes of Darwin and Marx is well placed, any attempt to downplay the passions and corporeal nature of man must also be rejected for they constitute a crucial part of man's nature.

Due to the luminosity provided by noetic insight, Plato and Aristotle were able to acknowledge their own ignorance. Indeed, both Plato and Aristotle place "wondering" as the starting point for philosophy.[43] In fact, Aristotle extends the dictum to everyone: "*All men by nature desire to know.*"[44] While all men experience this desire, it is stronger in some. The philosopher is one who is drawn by a sense of urgency to seek answers to the questions of his existence. Plato's parable of the Cave is the symbolization *par excellence* of this search.[45] The prisoner is moved to turn around (*periagoge*) and ascend out of the cave by a mysterious, unknown force.[46] The wisdom acquired outside of the cave is only attainable because of that force, but the draw alone is not enough: the prisoner must *respond* to the calling that compels him to turn around. It should be noted that the experience of wondering in Plato and Aristotle is not one of fear and anxiety: it is one of joy.[47] As Voegelin notes, this is "because the questioning has direction; the unrest is experienced as a theophanic event in which the *nous* reveals itself as the divine ordering force in which the *psyche* of the questioner and cosmos at large; it is an invitation to pursue its meaning into the actualization of noetic consciousness."[48]

Voegelin contrasts the classic experience of reason with some of the prominent representatives of modern reason in order to exemplify what happens when *nous* is disregarded and the directive nature of the search has been lost:

> In the modern Western history of unrest, on the contrary, from the Hobbesian "fear of death" to Heidegger's *Angst,* the tonality has shifted from joyful participation in a theophany to the *agnoia ptoiodes,* to the hostile alienation from a reality that rather hides than reveals itself. A Hobbes replaces the *summum bonum* with the *summum malum* as the ordering force of man's existence; a Hegel builds his state of alienation into a system and invites all men to become Hegelians; a Marx rejects the Aristotelian quest of the ground outright and invites you to join him, as a "socialist man," in his state of alienation, a Freud diagnoses the openness toward the ground as an "illusion," a "neurotic relict," and an "infantilism"; a Heidegger waits for a "parousia of being" which does not come, a Sartre feels "condemned to be free" and thrashes around in the creation of substitute meanings for the meaning he has missed.[49]

Yet, the shift represents more than just a change in tone or emphasis. The problem is that those thinkers invoke "reason" as justification for their respective systems. The *zoon noun echon* (the being that possesses nous) has been replaced by the *zoon agnoian echon* (the being that possesses ignorance).[50] But if *all* men desire by nature to know, then how can we make sense of this derailment? The answer lies within the *response* to the state of unrest. If man fails to utilize and nurture his highest capacity, noetic reasoning, then the questioning will have no direction. The sense of disorientation that results will then lead to artificial constructions of order, or second realities, in which the individual can gain a sense of certainty and control over his existence. Such constructions, however, must be protected because they will fail under rational scrutiny. Thus, a key tenet of modern systems, such as those offered by Marx and Comte, is to explicitly prohibit philosophical questioning in the classic sense.[51] A recovery of reason then will necessarily entail an "open" approach to politics: one that features philosophical questioning at its core.

Experience

Another term that has acquired a new meaning within modernity is *experience*. When faced with questions of verification, both classical and modern political philosophers will point to experience as proof, or evidence, of their respective claims. However, it is important to note the differences in what is meant by experience. For instance, when Bacon laments that none before him has "spent an adequate amount of time on things themselves and on experience,"[52] he is referring to observation and experimentation. Sense perception alone may mislead us, but the experiment serves as the "assistant to the senses" and allows us to gain true insight into nature. So while admitting the inadequacy of the senses if unaided, Bacon's "experience" is one based on information gained through the sense perception. This is a perfectly reasonable definition in relation to the natural sciences, but a crucial aspect of human existence is ignored if we simply import Bacon's notion of experience into the study of human affairs.[53]

Plato and Aristotle also appeal to experience, but they include an aspect that is missing in Bacon's account: *apperceptive* experience. This is an inner, participatory experience. As Ellis Sandoz notes in *The Voegelinian Revolution*, "it is, in any event, evident that the material and quantifiable do not exhaust the whole of experienced reality. No account that takes only these factors into consideration can form a sufficient basis for understanding man in his humanity or for understanding the science of politics."[54] Man's noetic capacity gives insight into areas of reality that are not exhausted by sensory perception. Insight into the structure of reality, and the realization of one's own participation in the *metaxy* are *apperceptive* experiences.[55] The divine draw, or pull, that man experiences, the force that compels him to turn around; these are just as real (Plato and Aristotle would argue that they are more so) as any sensory perception. The fact that man *participates* in this process precludes any notion of the traditional subject-object dichotomy that is prevalent in the methodology of natural science. Plato turns to the myth in part because *apperceptive* experiences are often ineffable. The medium of myth allows him to symbolize experiences that do not lend themselves easily to a standard, philosophical exegesis. This necessitates the understanding that politics is a relatively imprecise science for one can only be as precise as the subject matter allows.[56] And in addition to the questions raised by MacIntyre, man's *apperceptive* experiences are not expressible as mathematical formulae, nor can they be reduced to mere description.

Another important aspect of experience that is often neglected within modernity, and certainly in its scientistic manifestations, is common sense. Common sense serves as a line of defense against ideologies and systems that exclude or inhibit important aspects of political reality. As Thomas Reid notes:

> There is a certain degree of it which is necessary to our being subjects of law and government, capable of managing our own affairs, and answerable for our conduct towards others: This is called common sense, because it

is common to all men with whom we can transact business, or call to account for their conduct.[57]

Voegelin points out that it means the same as a "branch or degree of *ratio*" and represents the "habit of judgment and conduct of a man formed by *ratio*."[58] Voegelin points to Aristotle's *Politics* as an example of a "commonsense study," since it deals with situations that typically arise in society and history.[59] Common sense is also what allows us to recognize the deficiencies of ideological systems. So when Marx tells a hypothetical interlocutor to give up his questions about the nature of his existence or posits that man is a self-created being, common sense experience leads us to reject such propositions.[60] Marx's systematic ban on such questions serves as proof that he realizes the sway of common sense experience and therefore, he must find a way to convince his followers to ignore it.

Facts

Again, the problem of meaning arises in relation to the term "facts." As with reason and experience, most thinkers will claim to have "facts" on their side. And any science must rely on facts for its validation. Positivism helped initiate the separation between "fact" and "value." Under this conception, facts are held to be empirically demonstrable; that is to say, accessible through sense perception. Furthermore, facts cannot be tied to any sense of right or wrong; they are "value-free." To be scientific, on this account, is to engage in a discussion of facts. Thus, science is ethically neutral and the social scientist must strive to be so as well (to the extent he is being a good scientist). As Leo Strauss points out, such an attitude can lead not only to nihilism, but, perhaps more importantly, to a blind acceptance of the status quo.[61] As Strauss notes, "ethical neutrality . . . is not more than an alibi for thoughtlessness and vulgarity: by saying that democracy and truth are values, he says in effect that one does not have to think about the reasons why these things are good, and that he may bow as well as anyone else to the values that are adopted and respected by his society. Social science positivism fosters not so much nihilism as conformism and philistinism."[62] Strauss points to the almost unanimous acceptance of democracy amongst American social scientists as evidence of this claim. Furthermore, the social scientists in his homeland of Germany, offered little in the way of a critique of the Nazi regime, something that Eric Voegelin vigorously pointed out as well. That this was the case should not be too surprising given the positivistic outlook of those practitioners.

In stark contrast to the positive conception of social science, it is instructive to turn to Aristotle again. Far from being ethically neutral, Aristotle was adamant that ethics served as the foundation for political order.[63] Like the positive social scientist, Aristotle noted the importance of facts, but he was much more expansive in his definition. The student of politics has to study the soul first and foremost since the end of politics is the good of man.[64] The philosophical anthropology offered by Aristotle does not represent a "value judgment" or an "opinion." As Voegelin notes:

> Neither classical nor Christian ethics and politics contain value judgments but elaborate empirically and critically, the problems of order, which derive from philosophical anthropology as part of a general ontology. Only when ontology as a science was lost, and when consequently ethics and politics could no longer be understood as sciences of the order in which human nature reaches its maximal actualization, was it possible for this realm of knowledge to become suspect as a field of subjective, uncritical opinion.[65]

Man's nature includes not only the corporeal, but also the spiritual. The reductionist philosophical anthropologies offered by the likes of Comte, Marx, and Darwin fail to account for man's non-corporeal nature, and thus the systems that they base around such conceptions necessarily fail. They mistakenly subordinate theoretical relevance to methodology and exclude from examination the questions that matter the most in human affairs. Science is not to be characterized by any particular method. It provides insight into the nature and order of things. Thus, it aspires to rise above mere opinion and arrive at knowledge. The methods of the natural sciences are scientific only because they serve to provide insight into the subject matter that is being investigated, not because there is any inherent worth in the method itself. Yet, politics deals with the most complex of subjects, human beings, and as such, it cannot be adequately studied using methods that were designed to study different, and much simpler, subjects.[66] Moreover, ethics and politics are prudential sciences. They deal primarily with *action*. Because of this, rigid rules of action cannot be formulated. The fluidity of human affairs allows for generalizations and guideposts, but cannot be subjected to the rigorous quantification and precision found in fields with less complex subject matter. Facts must be taken to mean anything that corresponds to reality; a reality experienced by concrete human beings (whether it be of the perceptive or apperceptive variety). And thus a student of politics, far from being removed from politics, must be immersed in it.

Empiricism

Another term that deserves attention is empiricism. The meaning of the term has narrowed considerably within modernity, especially with the advent of positivism. Modern day social scientists often equate empiricism with measurability or with quantification. A study is said to be empirical if it utilizes objective, quantitative measures or if it refers to concrete historical cases. While such measures do indeed belong in the empirical realm, they are not exhaustive. The misunderstanding has arisen from the tendency to separate the *normative* from the *empirical*, with the assumption being that only the latter is scientific. The empirical merely states "what is" while the normative suggests "what ought to be." The problem with this distinction is similar to that found in the fact-value dichotomy: it unnecessarily reduces the legitimate realm of knowledge. *Any* study that bases its claims on experiential reality has the right to be called empirical.[67] An examination of human nature or political order is just as empirical as the number of representatives a state has in Congress. The standard for determining if a statement is empirical rests on its relation to reality and to the facts, not on its proclivity to quantification. Thus, the likes of Plato and Aristotle are not merely offering normative suggestions: they are conveying *empirical facts* about reality![68]

Science

In addition to the other terms mentioned thus far: reason, experience, facts, and empiricism, a change has also occurred in the meaning of science itself. Indeed, the problem of scientism is directly tied to the "new" meaning of science that has arisen within modernity. As noted throughout this study, science holds great prestige and any claim can be bolstered exponentially if it can be labeled as "scientific." Conversely, any claim that is seen as "unscientific" can readily be dismissed. Thus, the question about what constitutes science is paramount.[69] Without some idea of what constitutes science, we would be unable to distinguish pseudo-science from science and scientific claims would hold little persuasive power.

One aim that this study has in common with present day social science, and more specifically political science, is to be scientific. Yet this study is quite obviously "philosophical" so it is important to explain why it is also scientific. The distinction between philosophy and science is a modern invention. Prior to Bacon, philosophy and science were identical. "Natural philosophy" was the term applied to what we now refer to as natural science. Beginning with Bacon, the break with the Aristotelian tradition, led to a separation of philosophy and science. As Strauss explains:

> The distinction between philosophy and science or the separation of science from philosophy was a consequence of the revolution, which occurred in the seventeenth century. This revolution was primarily not the

victory of science over metaphysics, but what one may call the victory of the new philosophy or science over Aristotelian philosophy or science. Yet the new philosophy or science was not equally successful in all its parts. Its most successful part was physics (and mathematics). . . . The victory of the new physics led to the emergence of a physics which seemed to be as metaphysically neutral as, say, mathematics, medicine, or the art of shoemaking. The emergence of a metaphysically neutral physics made it possible for "science" to become independent of "philosophy" and in fact an authority for the latter. It paved the way . . . for the separation of political science from political philosophy as well as the separation of economics and sociology from political science.[70]

The new standard of science became physics and those in other fields tried their best to legitimate their studies by emulating it. The prestige afforded to the new science was something that all wished to share in. This is why Hume tried to imitate Newton's methodology in the field of ethics and why Comte went through great pains to create a "social physics."[71] With the rejection of Aristotle's physics came the rejection of his politics and ethics.[72] The problem arises from the fact that the methods that work so well in physics do not prove as fruitful in the realm of human affairs. And the misguided emphasis on method has led to an oversight of the problems that matter most.

An important aspect of the recovery of *nous* is the recognition that it is the driving force of what constitutes science in the first place. As Voegelin explains:

The truth of noesis has the character of rationality and science. . . . [I]t establishes the historical fact that the Platonic-Aristotelian noesis has developed the indices science (*episteme*) and theory (*theoria*) in order to express the character of their mode of knowledge and to distinguish it from the non-noetic modes of knowledge as the *doxai*, beliefs or opinions. "Science" is not determined by the mystery of a pre-existent definition, it rather discovers itself as the knowledge of the structure of reality, when consciousness historically attains the illumination of itself and its ratio. . . . Modern natural science is science not because we recognize it as such by convention but because the methods it uses to investigate the structure of the world are compatible with the *ratio* of noesis.[73]

Under this conception, the term "science" can indeed be appropriated beyond the methods of the natural sciences to the insights of Platonic and Aristotelian philosophy. And the assumption that "mathematical natural science" is the model of science itself is shown to be an "ideological dogma stemming from modern scientism."[74] The aspiration to be "scientific" in the narrow sense has unfortunately led social scientists to overlook the fundamental problems of human existence. The call for a new science of politics then is not simply a nostalgic desire to return to the past, but a necessity for the future. Just as political philosophy was concretely founded in response to corruption; so too is my study aimed directly at vitiating the deleterious effects of scientism.

Conclusion

In summary, the constructive features of a new science of politics can now be outlined. The first clue is in the very title of the chapter. A science of politics must at its core rest on a proper understanding of man. The good of man is the proper end of politics and that good cannot be determined or acquired unless man's full nature is accounted for. This means going beyond his corporeal nature and beyond the methods that are designed to study his phenomenal nature. As Voegelin points out, "A theory of politics must cover the problem of the order of man's *entire* existence."[75] This includes both the corporeal foundation and ordering consciousness; any attempt to focus on one or the other necessarily misses the mark.

Political science must again be conducted from the perspective of the citizen and statesman: not from an imaginary Archimedean point of objective neutrality. Political analysis must start from common sense experience and any attempt to systematize politics into a fixed set of propositions must be rejected. As a corollary, any analysis must ultimately be subjected to open questioning and the "proof" of a claim will be guaranteed not by a particular methodology but through experience, in the broad sense of the term. The search does not stop at the common sense level, however; science aspires to move beyond opinion into the realm of knowledge. Thus, reason becomes paramount and more specifically, *nous* must be accounted for. The "sensorium" for this capacity is consciousness and each man must account for his own existence within the *metaxy*.[76] Consciousness represents a *participatory* experience; the structure of reality is illuminated through the noetic capacity, but it also represents a process in the journey towards truth. In other words, noetic knowledge is *subjective*, but not solipsistic. The luminosity that Plato or Aristotle experiences is *representative* for all men and assuming that human nature is constant, it holds true throughout time.

Appeals to common sense experience and to consciousness may have an alienating effect on those who have been immersed in the positivist dogma and understandably so. Yet, those are indispensable parts of reality as experienced by man, and thus we cannot simply eliminate them because they do not happen to fit into the preferred way of doing things. As Sandoz succinctly summarizes, noetic science

> consists in the exploration of experience from the perspective of participation as its empirical basis; it sets aside the mathematizing method of the natural sciences and of conventional (or positivist) social science, as well as the extraneous language of subject-object, the dogmatic fact-value dichotomy, and the contraction of scientifically relevant "experience" to the world of sense perception as the controlling (if not sole) reality screening

scientific questions from pseudo-questions. Not to do so would be to abandon reality *as experienced*.[77]

Furthermore, if political science is to have any authority whatsoever in the realm of "real-world" politics, it must account for things that matter to the citizenry. While a glance at current scholarship would have us believe that the question of the best regime has been settled (democracy is clearly the answer), how are we to recognize when a not-so-innocuous form of government begins to arise? Does current political science possess the cognitive tools both to recognize disorder and to offer a corrective? The answer given by the likes of Voegelin, Strauss, Andreski, Hayek, MacIntyre, and Kolakowski is a resounding no, and this analysis confirms that conclusion. It is my contention that one of the key crises of our age, scientism, has infected both the political order and the very science that is designed to protect it. Just as the discovery of *nous* did not lead to a magical transformation of the political order in Athens, a new, or refined, political science cannot be expected to solve the perpetual problems of human existence. But it can at least allow us to recognize those problems and provide guidance on how to vitiate their effects.

Chapter 7

Conclusion: Scientism in the
Twenty-First Century and Beyond

"No one is obliged to take part in the spiritual crisis of a society; on the contrary, everyone is obliged to avoid this folly and live his life in order."[1]
–Eric Voegelin

While this analysis has encompassed a wide array of sources across hundreds of years, there is an underlying consistency in the manifestations of scientism. From Francis Bacon to present day, scientism has had a profound influence on politics, and there is little reason to believe that it will not continue to do so unless action is taken. The first step in any therapeutic endeavor is to properly identify the problem. Therefore, given the lack of systematic studies on the subject, I spent considerable time establishing scientism as the cause of significant political problems within modernity. From Bacon's dream of a scientific utopia in *New Atlantis* to Comte's religion of humanity to Marx's "scientific socialism" to the eugenics programs of the twentieth century, I have sought to demonstrate how scientism played an integral role in the formation of those politically dangerous and misguided endeavors. The therapy offered through a new science of politics is but a first step in what must be a long and arduous process if scientism's destructive effects are to be vitiated. While the problem is undoubtedly complex, there are several consistent features that have accompanied the scientistic attitude within modernity. These include a materialistic worldview, the diminishment of the *bios theoretikos,* the dogmatic faith in the *methods* of the natural sciences, the prohibition of philosophical questioning, and an emphasis on immanent fulfillment through the power of science. Perhaps the most troubling aspect is that scientistic programs are always presented as being *scientific*. Thus, the ability to distinguish between science and pseudo-science becomes paramount.

Even if this study has established the connection between scientism and the political problems outlined herein, it does little good if we do not possess the tools to recognize and prevent future manifestations. The recovery of reason, in the noetic sense, is the primary requirement for a new science of politics precisely because it provides us with the critical framework necessary to make political judgments. Noetic science provides insight into the structure of reality in addition to providing a model of order to aspire to. The recognition of disorder can come only if one has a proper model of order by which to critique it. The same holds for the distinction between science and pseudo-science. Pseudo-scientific claims can be recognized as manifestations of scientism only if our understanding of science is adequate. That understanding of science requires us to look beyond the methodology of the natural sciences and instead to focus on the *spirit* that motivates it. There is an inherent *openness* in the approach of a Newton or Einstein that can be likened to the "wondering" described by Aristotle and Plato as the proper beginning of philosophy, and subsequently of knowledge. And just as the philosophy of Plato and Aristotle suffered at the hands of subsequent dogmatizers, so too did Newton's rightly earned accolades lead to abuses by an appreciably influential group of followers.[2] Newton's insights came only because he was willing to step outside of conventional bounds. Had he bound himself to a particular methodology, he would have never revolutionized physics.[3] Likewise, the insights by Plato and Aristotle were epochal events in the history of man. They were just as scientific as Newton's, but their subject matter prevented them from offering the precision that Newton could provide in his *Principia*. Once a proper understanding of science is reached, the *scientistic* claims to authority lose much of their appeal since they are rooted in a highly restrictive, deformed view of science and reality.

Recognizing what constitutes legitimate science is only one of several steps that must be taken however. Given the increased control over nature that science has given man, we must be concerned with how that power is utilized. Indeed, the primary objective of the new science of politics outlined in the previous chapter is to provide practical guidance. The argument for a new science of politics is based first and foremost on the fact that contemporary political science has failed to adequately address the problem of scientism. What I find to be sorely lacking in modern social science, and specifically political science, is a fostering of *phronesis*, or practical wisdom. Politics is primarily a prudential science and thus requires an emphasis on how to act properly. An adequate science of politics then must be able to counsel the statesman in matters of concrete politics. This also requires an understanding of right and wrong and of justice: knowledge that is ultimately gained through the faculty of noesis.

Since it is my contention that the problem of scientism must be addressed by political science and that contemporary political science has been insufficient to the task, it is important to assess the prospects of a change in course within the discipline. From its inception, modern political science has aimed to provide the law-like generalizations found in the natural sciences. In the twentieth century, there was a concerted effort within American political science to move towards a more "scientific" understanding of politics. As John Shotwell wrote in 1921, "The only safety for democracy . . . is to apply scientific methods to the management of society as we have been learning to apply them in the natural world. . . . We are in the political sciences where the natural sciences were two hundred years ago."[4] And Bernard Crick, echoing contemporary critics such as Eric Voegelin and Leo Strauss, noted that the movement had only grown stronger by the time he published his seminal study, *The American Science of Politics,* in 1959.

Half a century later, the debate about the character and direction of political science is still not a settled issue. Political science has arguably become more "scientific" during that time, in the positivistic sense of the term. Methodology has even begun to gain acceptance as a subfield.[5] And some departments have recently decided to get rid of their political theory subfields altogether.[6]

Rehfeld raises concerns about the apparent confusion within the discipline and points to an "ongoing lack of engagement, suspicion, and animosity."[7] A primary reason for the discord lies precisely in the misconception of science. Few wish to engage in a discipline that is viewed as "unscientific" and if science is defined by the implementation of a particular method, then we can at least make sense of *why* political philosophy has fallen out of favor. It also explains why methodology has been conferred with so much prestige as to raise it to the level of a "substantive" field of inquiry. If progress is to occur, the foundation of the discipline must rest on the assumption that politics is a multi-faceted, complex subject matter. And as such, a variety of approaches are necessary to gain a sufficient understanding of that subject matter. Methodological thoroughness is indeed desirable, but theoretical *relevance* must be given priority.

Although scientism may manifest itself in various ways, there are several issues that will likely be of utmost concern in the near future. One issue previously discussed is eugenics, or genetic engineering. As knowledge of heredity and genetics has continued to increase, the prospects of human intervention has grown exponentially. The aforementioned Human Genome Project represents a sweeping, systematic inquiry into genetics that is likely to revolutionize the field. While the eugenics programs of the early twentieth century were modeled primarily on the enhancement of overall characteristics such as intelligence or physical strength, modern genetics has proven to be much more exact. We can now identify particular traits and predict with great accuracy, how those traits will be transmitted. It is no longer a question of *if* we will be able to control the transmission of certain traits, but *when*. Important ethical and political questions accompany such developments. Eugenics lost its appeal in the twentieth century largely due to the crude, and often inhumane, measures that were taken to enact the various programs devoted to it. Moreover, it was difficult to identify those who should be subjected to negative eugenics, making the programs inefficient in addition to being inhumane. But modern science provides us with the means to precisely identify genetic defects and in some cases, to correct those defects. Thus, the problem of inefficiency could conceivably be overcome, especially with further developments. And instead of forced sterilization, genetic engineering could provide a much more humane way to control the transmission of undesirable traits. It will be important to define what will and will not be allowed as technology continues to give us more control over nature and the answer to that question will lie, in part, on the view of man that predominates society.

Embryonic stem cell research is another "hot-button" political issue and just as with genetic engineering, the controversy is over the proper limits of scientific inquiry. Proponents of embryonic stem cell research point to the vast potential of the field to provide cures for debilitating diseases. This is because embryonic cells, unlike adult cells, can reproduce themselves indefinitely and take the form of any type of cell. At least in theory, such cells could be used to cure a variety of afflictions such as Parkinson's disease, cancer, and spinal cord injuries. Opponents are quick to point out that practical benefits have yet to occur and more importantly, the research requires the destruction of human embryos. As Peter Boyer points out, even the originator of the process, Dr. James Thomson, expressed concerns: "If human embryonic stem cell research does not make you at least a little uncomfortable, you have not thought about it enough."[8] President George W. Bush agreed and placed restrictions on embryonic stem cell research. Federal funding was restricted and the lines that had been previously established could be utilized, but no new lines could be created. As part of his 2008 campaign, Barack Obama had pledged to support embryonic stem cell research and one can infer that he had it in mind in his inaugural address when he pledged to "restore science to its rightful place." Just two months after his inauguration, President Obama made good on that promise by issuing an executive order that lifted the restrictions imposed by President Bush.[9] President Obama's decision was lauded by political supporters and also had appreciable support within the scientific community, but it was recently halted by a preliminary injunction issued in a United States district court.[10] The issue ultimately hinges on a delicate ethical decision, but the rhetoric has often been vehement, as some proponents of embryonic stem cell research have accused opponents of being "anti-science." And since those opponents often use their religious faith as justification for their positions, the issue has been framed as one of science against religion.

Part of my argument has been that the current worldview is one infected by scientism. The view of man and his place in the world is still largely governed by Darwinian concepts. This is evident in everything from the debates on school curriculum to the place of religion within society. Invoking Darwin as authority, the New Atheists, such as Daniel Dennett, Sam Harris, Richard Dawkins and the late Christopher Hitchens, have staunchly argued against religion, in the name of science, humanity and reason.[11] Dawkins, an Oxford trained biologist, is one of Darwin's most outspoken supporters, so much so that he earned the nickname "Darwin's Rottweiler."[12] Dawkins considers the evidence of evolution to be overwhelming and contends that religion and evolution are strictly incompatible. Just as Darwin had admitted in his autobiography that his theory of natural selection undermined his former beliefs of design in the universe, Dawkins adamantly rejects any explanation that encompasses a divine creator. The primary difference is that Darwin did not claim to refute the idea with certainty, while Dawkins does. Darwin's work, as noted previously, does not speak to questions of a first cause or the origin of life itself. He simply explains how life, once given, evolves. Yet Dawkins takes it to be a settled fact that evolution did not require a master craftsmen or "watchmaker." He therefore chides those both within and outside of the scientific community who try to find compatibility between evolution and any type of creationism. In fact, any appeal to the supernatural is strictly off limits.[13] As Dawkins forcefully states, "I am not attacking any particular version of God or gods. I am attacking God, all gods, anything and everything supernatural, wherever and whenever they have been or will be invented."[14]

Dawkins' approach is clearly *scientistic*, as he places a dogmatic faith in science and takes his argument much further than the evidence allows. His foundation, The Richard Dawkins Foundation for Reason and Science, is dedicated to expunging religion from society under the prestigious cover of science.[15] However, Dawkins is not simply an outlier or the representative of a fringe movement, as evident by the millions of copies that have been sold of *The God Delusion*. He is adamantly opposed to any worldview that encompasses the transcendent or spiritual nature of man. The reductionist, biological view of man he presents is particularly dangerous because he frames it as scientific.[16] And nobody wishes to argue with science. Anyone opposed to Dawkins then must be anti-scientific and irrational.

Sam Harris, who has a background in neuroscience, similarly attacks religion in the name of science. He is the co-founder and CEO of Project Reason, which is "devoted to spreading scientific knowledge and secular values in society. The foundation draws on the talents of prominent and creative thinkers in a wide range of disciplines to encourage critical thinking and erode the influence of dogmatism, superstition, and bigotry in our world."[17] In his bestselling *The End of Faith: Religion, Terror and the Future of Reason*, Harris, inspired by the events of 9/11, offers harsh and widespread criticisms of both Islam and Christianity. Harris does not stop at merely criticizing religion as fundamentally flawed: he tries to provide an adequate substitute. In *The Moral Landscape,* Harris argues that science can provide guidance in the moral realm. Stepping outside of the modern division between natural science and morality, Harris thinks science can provide us with a universal morality and answer questions about right and wrong. This is a response to the charge by many believers, and some non- believers, that society would fall apart if religion did not provide some sense of morality. In other words, whether it is true or not, they argue religion can be useful in providing order. Harris thinks that religion has hurt society much more than it has helped and looks to science to provide the moral guidance formerly offered, albeit inadequately, by religion.

I did not list atheism as a central feature of scientism, although it is somewhat implied in the radically immanent worldview. Part of the reason for this omission is because atheism is often understood as being anti-religious and in fact, those such as Dawkins often couch their own arguments as being so. But scientism *is* religious in the sense that its adherents express a dogmatic faith in its power. Dawkins expresses the type of certainty that is not found within science or philosophy properly speaking; he speaks as forcefully as the most religious of the "zealots" he criticizes. The *faith* that Dawkins places in science is overlooked because he has rhetorically placed faith in the realm of religion. This brings us back to the importance of language and exemplifies why the recovery of the original meanings of "reason" and "science" is vital. Reason, in the noetic sense, is not antithetical to faith (indeed it relies on it) and is grounded in the understanding of man's participatory experience in the metaxy. This encompasses the corporeal drive of the passions, but it also includes the "divine" pull that draws us towards truth. Because of that component, it can be argued that the discovery of *Nous* is itself a revelatory event.[18] And science represents knowledge about the nature of things. It is not limited to phenomenal relations or to a particular methodology and is characterized by an inherent openness to reality. If we understand science and reason in this manner, then Dawkins' argument is exposed as both antithetical to reason and science. This is because it obliterates the highest capacity of the human intellect, *nous*, and unnecessarily reduces man's nature to the corporeal. Dawkins' scientific credentials are impressive, but to the extent he makes arguments such as those found in *The God Delusion*, he is doing so not as a scientist, but as a purveyor of *scientism*.

A final area that is undoubtedly prone to scientistic infection is climate change and environmental policy. As briefly mentioned previously, Al Gore's *An Inconvenient Truth* helped draw popular attention to the issue of climate change. Rising oil prices and devastating natural disasters, such as Hurricane Katrina, further increased the saliency of the issue as some asserted that the events were essentially self-inflicted. Gore, much like Dawkins, appeals primarily to science as justification for his claims. And to be sure, evidence does exist that the planet's temperature has been increasing over the past century. That rise in temperature has roughly coincided with the industrialization of large population centers, such as China and India. Gore takes such a correlation to be proof of causation, so much so that he calls human made global warming an "absolutely indisputable fact."[19]

The problem is that not all scientists agree, which of course means that the fact is not quite as certain as Gore presents it. A quick glance at any geology textbook will show that climate change is a constant phenomenon in the history of the earth. According to geologists, there have been several periods with much warmer temperatures than anything humans have experienced; just as there have been much cooler ages. This general fact seems to point to the possibility that recent increases in the earth's temperature may have occurred to some extent, regardless of human intervention. Furthermore, recent data shows that the earth's temperature has not been climbing as quickly as predicted by Gore.[20] The problem with the whole ordeal is that the debate is being conducted with political motivations. Proponents of climate change have a vested interest in science confirming their position.[21] And likewise, there are those who benefit from the status quo, such as those within the traditional energy sector, and have an interest in denying, or minimizing, the negative environmental effects of their respective industries. Both sides realize that science can be used as a powerful tool to protect their political interests. And they are willing to ignore evidence that undermines their positions. This is a lingering effect of scientism that is likely to continue well into the future.

As a last point of emphasis, I wish to clarify both the intentions and implications of the study. Clearly, scientism has been identified as a political and intellectual problem of utmost importance. Although the proposed solution encompasses a return to the principles of classical political science, I do not wish to suggest that its content be directly imported.[22] The necessity of a return to those principles is due to the dogmatic closure to fundamental aspects of reality. Science and philosophy, properly speaking, are open endeavors that are motivated by the search for truth about the nature of things. Therefore, the "answers" are not necessarily as important as the asking of the Question.[23] Each individual, as well as each society, must reflect upon the pressing questions that are necessitated by the participation in the *metaxy*, or reality. A key feature of scientism is the prohibition of such questions, and thus my solution embraces an approach that not only allows, but also encourages such questions. Likewise, an incessant focus upon a particular methodology represents a closure to important aspects of reality; deforming science in principle and leading to less than desirable political consequences in practice. In regards to natural science, it can undoubtedly continue to provide us with helpful innovations. Advances in medicine and technology will, and should, continue to flourish. Problems do not arise from science itself, but from men who wish to use its power to dominate others. Therefore, what is needed is a comprehensive science of man: one that can provide ethical and political guidance.

I began the study with President Obama's assertion that we must "restore science to its rightful place." This implied that science once was in its rightful place and that somehow it had been derailed. To anyone that followed Obama's campaign, it is clear he meant to suggest science had been derailed by his predecessor, George W. Bush. By coming out against stem cell research and apparently not taking climate change seriously, Bush was turning a blind eye to science.[24] Obama promised to bring science back to prominence, which presumably meant that he would let it guide policy decisions or at least form an integral role in shaping them. While I agree with the assertion that science should be put in its rightful place, the task is necessary not because of the policies of any particular leader or lawmaker, but primarily because of the pernicious effects of scientism. Restoring science to its rightful place requires us to recover the original meaning of science, and in regards to natural science, we must recognize the inherent limits to its guidance and the dangers of its misapplication. Natural science cannot tell us how to act or which ends to pursue. Instead, it aspires to tell us how things work within the natural world. And as Bacon rightly pointed out, such knowledge can lead to practical utility by giving us power over nature. But it does not provide us with guidance as to how that power should be utilized and as shown throughout the study, it can be used to destroy man just as easily as it can to help his cause. Thus, what is truly needed is a *political* science: one that can speak to questions of the human good. The rise of natural science has given man unprecedented power over nature, but without a comprehensive science of human affairs, abuses of that power are likely to continue. Therefore, my project aspires to restore political science to its rightful place as the master science of human affairs. And in so doing, it represents one step in the arduous effort required to restore science to its rightful place.

Bibliography

Andreski, Stanislav. *Social Sciences as Sorcery*. New York: St. Martin's Press, 1972.

Aristotle. *Nicomachean Ethics*. Translated by Martin Oswald. Upper Saddle River, NJ: Prentice Hall, 1999.

———. *Politics*. Translated by H. Rackham. Cambridge, MA: Harvard University Press, 1932.

Bacon, Francis. *Essays and New Atlantis*. Edited by Gordon Haight. New York: Walter Black Publishing, 1942.

———. *The Advancement of Learning*. Edited by Kitchin. London and Melbourne: Dent Publishing, 1973.

———. *New Atlantis and The Great Instauration*. Edited by Jerry Weinberger. Illinois: Harlan Davidson, 1989.

———. *The New Organon [Novum Organum]*. Edited by Jardine and Silverthorne. Cambridge: Cambridge University Press, 2000.

Boyer, Peter. "The Covenant." *The New Yorker*, September 6, 2010.

Brooke, John. "Darwin and Victorian Christianity." In *Cambridge Companion to Darwin*, 2nd edition, edited by Hodge and Radick. Cambridge: Cambridge University Press, 2009.

Burleigh, Michael. *Earthly Powers: The Clash of Religion and Politics in Europe from the French Revolution to the Great War*. New York: Harper Collins, 2005.

Ceaser, James. "The Roots of Obama Worship." *Weekly Standard*, January 25, 2010.

Clark, Gordon. *The Philosophy of Science and Belief in God*. Trinity Foundation, 1964.

Cohen, I. Bernard. "The Case of the Missing Author: The Title Page of Newton's *Opticks* (1704)." In *Isaac Newton's Natural Philosophy*, edited by Jed Z. Buchwald and I. Bernard Cohen. Cambridge, MA: MIT Press, 2004.

Comte, Auguste. *Introduction to Positive Philosophy*. Edited by Frederick Ferre. Indianapolis: Hackett Publishing, 1998.

———. *System of Positive Polity*. Translated by John Henry Bridges. London: Longmans, Green, and Co., 1875.

———. *The Catechism of Positive Religion*. Translated by Richard Congreve. 3rd edition. London: Ballantyne, Hanson and Co. Printing, 1891.

———. *The Essential Comte*. Edited by Stanislav Andreski. Translated by Margaret Clarke. London: Croom Helm, 1974.

Coetsier, Meins. "God the Creative Ground of Existence in Voegelin, Etty Hillesum and Martin Buber: A Response to Richard Dawkins' *The God Delusion*." Conference paper given at the annual meeting of the Eric Voegelin Society, Boston, MA, 2008.

Crick, Bernard. *The American Science of Politics*. London: Routledge and Kegan Paul Publishing, 1959.

D'Alembert, Jean Le Rond. *Preliminary Discourse to the Encyclopedia*. Edited by Richard Schwab. Chicago: University of Chicago Press, 1995.

Darwin, Charles. *On the Origin of the Species*. New York: Penguin Books, 2009.

———. *The Descent of Man*. New York: Penguin Books, 2004.

Dawkins, Richard. *The Blind Watchmaker: Why the Evidence of Evolution Reveals a World without Design*. New York and London: W.W. Norton and Company, 1996.

———. *The God Delusion*. Boston and New York: Mariner Books, 2006.

De Mesquita, Bueno. *The War Trap*. New Haven and London: Yale University Press, 1984.

Dewey, John. *Reconstruction in Philosophy*. New York: Henry Holt and Co., 1950.

Feyerabend, Paul. *Against Method*. 3rd edition. New York and London: Verso, 1993.

Gartzke, Eric. "War is in the Error Term." *International Organization* 53 (1999): 567–87.

Gore, Albert. *An Inconvenient Truth: The Planetary Emergency of Global Warming and What We Can Do About It*. New York: Rodale Publishing, 2006.

Gould, Stephen Jay. *The Mismeasure of Man*. Revised and Expanded Edition. New York and London: W.W. Norton, 1996.

Hall, Stephen. "Darwin's Rottweiler." *Discover Magazine*, September 2005.

Hallowell, John. *Main Currents in Modern Political Thought*. New York: Henry Holt and Co., 1950.

Harris, Sam. *The End of Faith: Religion, Terror, and the Future of Reason*. New York: W.W. Norton and Co., 2005.

Hayek, F. A. *The Counter-Revolution of Science: Studies on the Abuse of Reason*. Indianapolis: Liberty Press, 1952.

Hobbes, Thomas. *Leviathan*. Oxford and New York: Oxford University Press, 1998.

Hume, David. *A Treatise of Human Nature*. Edited by L.A. Sigby-Bigge. Oxford: Clarendon Press, 1896.

Husserl, Edmund. *The Crisis of European Sciences and Transcendental Phenomenology*. Translated by David Carr. Evanston, IL: Northwestern University Press, 1970.

Kennington, Richard. *On Modern Origins: Essays in Early Modern Philosophy*. Edited by Kraus and Hunt. Lanham, MD: Lexington Books, 2004.

Kitchener, Philip. "Giving Darwin His Due." In *Cambridge Companion to Darwin*, 2nd edition, edited by Hodge and Radick. Cambridge: Cambridge University Press, 2009.

Kolakowski, Leszek. *Main Currents of Marxism.* New York: W.W. Norton and Company, 2005.

———. *The Alienation of Reason: A History of Positivist Thought.* Translated by Norbert Guterman. New York: Anchor Books, 1969.

Kuhn, Thomas. *The Structure of Scientific Revolutions.* 3rd edition. Chicago: University of Chicago Press, 1996.

Lomborg, Bjorn. *The Skeptical Environmentalist: Measure the Real State of the World.* Cambridge, UK: Cambridge University Press, 2001.

Lyon, Peyton. "Saint-Simon and the Origins of Scientism and Historicism." *The Canadian Journal of Economics and Political Science* 27.1 (1961): 55–63.

Machiavelli, Niccolo. *The Prince.* 2nd edition. Translated by Mansfield. Chicago: University of Chicago Press, 1998.

MacIntyre, Alasdair. *After Virtue.* 2nd edition. Notre Dame, IN: University of Notre Dame Press, 1984.

Marx, Karl. *Economic and Philosophic Manuscripts of 1844.* Translated by Milligan. New York: Prometheus Books, 1988.

Marx, Karl and Frederick Engels. *Communist Manifesto.* Translated by Milligan. New York: Prometheus Books, 1988.

McKnight, Stephen. *The Modern Age and the Recovery of Ancient Wisdom.* Columbia, MO: University of Missouri Press, 1991.

———. *The Religious Foundations of Francis Bacon's Thought.* Columbia, MO: University of Missouri Press, 2006.

Newton, Isaac. *Philosophical Writings.* Edited by Janiak. Cambridge: Cambridge University Press, 2004.

Oakeshott, Michael. *Rationalism in Politics and Other Essays.* New and Expanded Edition. Indianapolis: Liberty Fund, 1991.

Paul, Diane. "Darwin, Social Darwinism, and Eugenics." In *Cambridge Companion to Darwin,* 2nd edition, edited by Hodge and Radick. Cambridge: Cambridge University Press, 2009.

Pickering, Mary. *Auguste Comte: An Intellectual Biography.* Volume 1. Cambridge: Cambridge University Press, 1993.

Plato. *The Great Dialogues of Plato.* Translated by W.H.D. Rouse. New York: Signet Classic, Penguin Books, 1999.

———. *Theaetetus.* Translated by Sachs. Newbury Port, MA: Focus Publishing, 2004.

———. *Timaeus and Critias.* Translated by Jowett. Scotts Valley, CA: IAP, 2009.

Polanyi, Michael. *Personal Knowledge: Towards a Post-Critical Philosophy.* Chicago: University of Chicago Press, 1974.

Rehfeld, Andrew. "Offensive Political Theory." *Perspective on Politics* 8 (2010): 465–86.

Rossi, Paolo. *Francis Bacon: From Magic to Science.* Chicago: University of Chicago Press, 1968.

Rousseau, Jean Jacque. *The Discourses and Other Early Political Writings.* Edited by Victor Gourevitch. Cambridge: Cambridge University Press, 1997.

Rummel, R. J. *Death by Government.* New Jersey: Transaction Publishers, 1994.

Sandoz, Ellis. *The Voegelinian Revolution.* 2nd edition. New Jersey: Transaction Publishers, 2000.

Snow, C.P. *The Two Cultures.* Edited by Collini. Cambridge: Cambridge University Press, 1998.

Stoner, James R. "Politics and Science." *Public Discourse: Ethics Law and the Common Good.* March 20, 2009. http://www.thepublicdiscourse.com/2009/03/77/.

Strauss, Leo. *Liberalism: Ancient and Modern.* Chicago and London: University of Chicago Press, 1968.

———. *What Is Political Philosophy? And Other Studies.* Chicago: University of Chicago Press, 1988.

Voegelin, Eric. *Anamnesis.* Translated by Gerhart Niemeyer. Columbia, MO: University of Missouri Press, 1978.

———. *From Enlightenment to Revolution*. Durham, NC: Duke University Press, 1975.

———. *Science, Politics, and Gnosticism*. Edited by Sandoz. Wilmington: ISI Books, 2004.

———. *The English Quest for the Concrete*. Volume 24 of the Collected Works of Eric Voegelin. History of Political Ideas, Volume VI: *Revolution and the New Science*. Edited by Barry Cooper. Columbia, MO: University of Missouri Press, 1998.

———. *The History of the Race Idea: From Ray to Carus*. Volume 3 of the Collected Works of Eric Voegelin. Edited by Vondung. Columbia, MO: University of Missouri Press, 1998.

———. *The New Order and Last Orientation*. Volume 25 of the Collected Works of Eric Voegelin. Edited by Gebhardt and Hollweck. Columbia, MO: University of Missouri Press, 1999.

———. *The New Science of Politics: An Introduction*. Chicago and London: University of Chicago Press, 1952.

Wernick, Andrew. *Auguste Comte and the Religion of Humanity*. Cambridge: Cambridge University Press, 2001.

West, John. *Darwin Day in America*. Wilmington, DE: ISI Books, 2007.

White, Howard. *Peace Among the Willows: The Political Philosophy of Francis Bacon*. The Hague: Martinus Nijhoff Press, 1968.

White, Michael. *Isaac Newton: The Last Sorcerer*. Reading, MA: Perseus Books, 1997.

Wilson, Woodrow. Constitutional Government in the Papers of Woodrow Wilson. Edited by Arthur Link. Princeton, NJ: Princeton University Press, 1974.

———. *The New Freedom*. Edited by William Leuchtenburg. Englewood Cliffs, NJ: Prentice Hall, 1961.

Endnotes

Scientism and the Dogmatics of Modernity

1   Eric Voegelin, The English Quest for the Concrete, Volume VI: Revolution and the New Science, ed. Barry Cooper, Collected Works of Eric Voegelin: History of Political Ideas 24 (Columbia, MO: University of Missouri Press, 1998), 214–15.

2   http://www.nytimes.com/2009/01/20/us/politics/20text-obama.html.

3   R. J. Rummel has coined the term "democide" to refer to the killing of citizens by their own governments. Democide accounted for far more deaths than war in the twentieth century (including 62 million in the Soviet Union alone). See R. J. Rummel, *Death by Government* (New Brunswick, NJ: Transaction Publishers, 1994), 24.

4   For an excellent study on the race idea see Eric Voegelin's *The History of the Race Idea: From Ray to Carus*, ed. Klaus Vondung, Collected Works of Eric Voegelin 3 (Columbia, MO: University of Missouri Press, 1998)

5   I am especially indebted to the work of Eric Voegelin, both in my analysis of scientism and in the solution offered through a new science of politics.

6   Eric Voegelin, *The New Science of Politics: An Introduction* (Chicago and London: University of Chicago Press, 1952), 4.

7   Ibid., 4–5

8   Leszek Kolakowski, *The Alienation of Reason: A History of Positivist Thought*, trans. Norbert Guterman (New York: Anchor Books, 1969), 204.

9   Ibid., 206.

10 Ibid., 208.

11 F. A. Hayek, The Counter-Revolution of Science: Studies on the Abuse of Reason (Indianapolis: Liberty Press, 1952), 21.

12 Such as Auguste Comte and Francis Bacon. Neither had a noteworthy scientific achievement. Hayek calls Bacon the "demagogue of science" (ibid., 21).

13 Ibid., 22.

14 Ibid., 24.

15 Ibid.

16 Leo Strauss, *Liberalism: Ancient and Modern* (Chicago and London: University of Chicago Press, 1968), 211.

17 Ibid., 212.

18 Ibid., 212.

19 Ibid., 214.

20 Ibid., 214.

21 Michael Oakeshott, *Rationalism in Politics and Other Essays*, New and Expanded Edition (Indianapolis: Liberty Fund, 1991), xvi.

22 Ibid., 9.

23 Ibid., 12.

24 Ibid.

25 Ibid., 13.

26 Voegelin takes this to be the primary task of the political philosopher as exemplified concretely in the person of Socrates.

27 Edmund Husserl is an obvious exception to this, but his solution of founding an "apodictic philosophy" is less than desirable. See Edmund Husserl, *The Crisis of European Sciences and Transcendental Phenomenology*, trans. David Carr (Evanston, IL: Northwestern University Press, 1970), 72.

28 Eric Voegelin, *The English Quest for the Concrete, Vol. 6: Revolution and the New Science*, ed. Barry Cooper, Collected Works of Eric Voegelin: History of Political Ideas 24 (Columbia, MO: University of Missouri Press, 1998), 208.

29 Although the same cannot be said of the founders of modern natural science. For example, Isaac Newton took alchemy quite seriously and spent considerable time trying to harness powers, which would today be described as magical, from it. See Michael White, *Isaac Newton: The Last Sorcerer* (Reading, MA: Perseus Books, 1997).

30 Or at the least, comes about shortly after its inception with the philosophy of Francis Bacon.

31 See Richards Schwab's introduction to Jean Le Rond D'Alembert, *Preliminary Discourse to the Encyclopedia* (Chicago: University of Chicago Press, 1995), xxv.

32 See Peyton Lyon, "Saint-Simon and the Origins of Scientism and Historicism," *The Canadian Journal of Economics and Political Science* 27.1 (1961): 62.

33 Andrew Wernick, *Auguste Comte and the Religion of Humanity* (Cambridge: Cambridge University Press, 2001), 3.

34 Ibid., 5.

35 Auguste Comte, *Introduction to Positive Philosophy*, ed. Frederick Ferre (Indianapolis: Hackett Publishing, 1998), 29.

36 Ibid., 17.

37 Ibid., 18.

38 Wernick, *Auguste Comte*, 185.

39 "Our sole object here is to effect, for the new religion, a general exposition equivalent to that which formerly taught you Catholicism. This second operation ought to be even easier than the former, for not only is your reason now mature, but the doctrine is by its nature, more intelligible as always demonstrable" (Auguste Comte, *The Catechism of Positive Religion*, trans. Richard Congreve, 3rd ed. [London: Ballantyne, Hanson and Co. Printing, 1891], 49).

40 Comte, *Catechism*, 53.

41 Karl Marx, *Economic and Philosophic Manuscripts of 1844*, trans. Martin Milligan (New York: Prometheus Books, 1988), 110.

42 Ibid., 110.

43 Ibid.

44 Ibid., 111.

45 Ibid.

46 Ibid.

47 Ibid.

48 Karl Marx and Frederick Engels, *Communist Manifesto*, trans. Martin Milligan (New York: Prometheus Books, 1988), 243 (emphasis added).

49 John West, *Darwin Day in America* (Wilmington, DE: ISI Books, 2007), 34.

50 John Brooke, "Darwin and Victorian Christianity," in *Cambridge Companion to Darwin,* ed. Jonathan Hodge and Gregory Radick, 2nd ed. (Cambridge: Cambridge University Press, 2009), 201.

51 West, Darwin Day, 38.

52 Charles Darwin, *The Descent of Man* (New York: Penguin Books, 2004), 679.

53 West, *Darwin Day*, 26–27. See Darwin, *Descent of Man*, 100.

54 West, *Darwin Day*, 27. See Darwin, *Descent of Man*, 110.

55 West, *Darwin Day*, 28. See Darwin, *Descent of Man*, 682.

56 Darwin, *Descent of Man*, 116.

57 "The idea of a universal and beneficent Creator of the universe does not seem to arise in the mind of man, until he has been elevated by long-continued culture" (Darwin, *Descent of Man*, 682).

58 See Diane Pauls, "Darwin, Social Darwinism, and Eugenics," in *Cambridge Companion to Darwin*, 234. In addition to Galton, Darwin's son Leonard was a strong supporter of eugenics as evident in his role of President of the Eugenics Society in Britain (See Pauls, "Darwin," 241). Thus while Darwin himself never fully embraced eugenics, there is ample evidence that his system led to its implementation.

59 West, *Darwin Day*, 139. According to Kelves, it jumped fivefold from 2–4 per one hundred thousand to 20 per hundred thousand from 1920 to the end of the 1930's.

60 Voegelin, *History of the Race Idea*, 23.

61 Ibid., 3.

62 Ibid., 3–4.

63 I rely heavily on Voegelin's notion of a "noetic science." For a good synopsis, see Ellis Sandoz, "*Principia Noetica*," in *The Voegelinian Revolution*, 2nd ed. (New Jersey: Transaction Publishers, 2000), 188–216.

Francis Bacon and the New Science

1 Abraham Cowley's preface to Thomas Sprat's early *History of the Royal Society*.

2 Locke and Newton were the other two. See Jefferson to John Trumbull, February 5th, 1789, in *Jefferson: Writings* ed. Merrill Peterson (Literary Classics of the U.S., 1984), 939.

3 Jean Jacques Rousseau, *The Discourses and Other Early Political Writings*, ed. Victor Gourevitch (Cambridge: Cambridge University Press, 1997), 332.

4 John Dewey, *Reconstruction in Philosophy* (New York: Henry Holt and Co., 1950), 46.

5 Paolo Rossi, *Francis Bacon: From Magic to Science* (Chicago: University of Chicago Press, 1968), xiii.

6 See Lisa Jardine's introduction to Francis Bacon, *The New Organon* (*Novum Organum*), ed. Lisa Jardine and Michael Silverthorne (Cambridge: Cambridge University Press, 2000), xv.

7   For example, the rejection of Aristotle's physics need not lead to the rejection of his politics. See Alasdair MacIntyre, *After Virtue*, 2nd ed. (Notre Dame, IN: University of Notre Dame Press, 1984).

8   Novum Organum, 98.

9   Experimentation on humans is especially problematic for two reasons. One is due to epistemological reflexivity. Those being observed know they are being observed and thus may react differently than in "natural" conditions. Secondly, there are serious ethical constraints as to what kind of experimentation can be implemented.

10 I deal with the question of "substance" at length in Chapter 5.

11 Francis Bacon, *The Advancement of Learning*, ed. G. W. Kitchin (London and Melbourne: Dent Publishing, 1973), 37.

12 This is not an altogether novel claim. While the likes of Dewey, Jefferson, and Rousseau heaped praise upon Bacon, F.A. Hayek referred to him as the "demagogue of science." See F. A. Hayek, *The Counter-Revolution of Science: Studies on the Abuse of Reason*, 2nd ed. (Indianapolis: Liberty Press, 1979), 21.

13 Howard White, *Peace Among the Willows: The Political Philosophy of Francis Bacon* (The Hague: Martinus Nijhoff Press, 1968), 126.

14 Rossi, *Francis Bacon*, 102. Rossi's interpretation diverges from White's. Rossi argues Bacon was heavily influenced by alchemy and sought to recover ancient knowledge. He is not a modern in that sense.

15 White designates Bacon as the first "modern utopian." Bacon exhibits faith that the unknown will be conquered via science. This is in contrast to ancient utopianism (which was actually realistic according to White). In the utopias of More and Plato, science serves virtue and remains the handmaiden of philosophical wisdom. See White, *Peace Among the Willows*, 97–99.

16 Novum Organum, xviii.

17 Ibid.

18 Ibid., 11.

19 Advancement of Learning, 205.

20 See Jerry Weinberger's introduction to *New Atlantis*. "The full importance of Bacon's teachings about the new science can be seen only when the reader realizes that the center of that teaching is a 'secret and retired' political science" (Weinberger, Introduction to Francis Bacon, *New Atlantis and The Great Instauration*, ed. Jerry

Weinberger [Arlington Heights, IL: Harlan Davidson, 1989], xii).
This implies that politics is too dangerous to discuss openly.

21   Rossi, *Francis Bacon*, 85.

22   Rossi, *Francis Bacon*, 86.

23   This applies especially to the Anglican Church, although he took
      precautions in offending the Catholic Church as well. For example,
      he took out some controversial statements from the original Ad-
      vancement of Learning when publishing the second edition.

24   These issues will be discussed in detail in the exposition on *New At-
      lantis*.

25   Novum Organum, 15.

26   Ibid., 15–16.

27   Ibid., 16.

28   Ibid., 8.

29   Ibid., 11.

30   Ibid., 18.

31   Ibid., 98–99.

32   Ibid., 64.

33   Ibid., 40.

34   Ibid., 41.

35   Ibid., 46.

36   Ibid., 48.

37   Ibid., 49.

38   Ibid.

39   Most notably, he includes Aristotle, Plato, and the Scholastics. See
      Ibid., 51–53.

40   Ibid., 56.

41   Ibid., 76.

42   Ibid., 77.

43   Ibid., 78.

44   Ibid.

45   Ibid., 79–80.

46   Ibid., 81.

47   Ibid., 86.

48   Ibid., 88 (emphasis added).

49   Ibid.

50   Ibid.

51   Ibid., 72.

52   Ibid., 24.

53 It should be noted that the *New Atlantis* was not published until after his death. So in a literal sense, he can be absolved of the charge.

54 Weinberger, Introduction, xii.

55 Ibid., xiii. While this claim may stretch the evidence, Weinberger's overall assertion that the work is crucial to Bacon's project is well founded.

56 See Plato, *Timaeus and Critias*, trans. Benjamin Jowett (Scotts Valley, CA: IAP, 2009).

57 Weinberger, Introduction, xiv. A speech that presumably would have been about why the Atlantians were to be punished: their lack of moderation.

58 Novum Organum, 24.

59 White, Peace Among the Willows, 94.

60 Weinberger suggests that those unwilling or unfit to stay "must have been restrained by force or killed" (Weinberger, Introduction, xvi).

61 Francis Bacon, *Essays and New Atlantis*, ed. Gordon Haight (New York: Walter Black Publishing, 1942), 254.

62 In earlier works, Bacon had often insisted "that we do not place our felicity in knowledge, as we forget our mortality" (*Advancement of Learning*, 6).

63 New Atlantis, 260.

64 Ibid., 268.

65 Ibid., 269.

66 See White's *Peace Among the Willows* and Weinberger's introduction to *New Atlantis*.

67 Stephen McKnight, *The Religious Foundations of Francis Bacon's Thought* (Columbia, MO: University of Missouri Press, 2006), 15.

68 Ibid., 3.

69 Ibid.

70 Ibid.

71 McKnight is right to emphasis the religious character of Bacon's work, but the evidence seems to suggest that science, not Christ, is what is to be worshipped. White and Weinberger are justified in questioning Bacon's fidelity to Christianity, although perhaps not to religion in general.

72 Consider one such passage: "We answered, after we had looked awhile one upon another, admiring this gracious and *parent-like* usage, that we could not tell what to say, for we wanted words to express our thanks; and his noble free offer left us nothing to ask. It

seemed to us that we had before us a picture of our *salvation in heaven*" (*New Atlantis*, 255–56; emphasis added).

73   Ibid., 274.
74   Weinberger, Introduction, xxiv.
75   Ibid., 65.
76   Ibid., 66.
77   Ibid., 66–67.
78   Ibid., 68.
79   White, Peace Among the Willows, 139.
80· Ibid., 187.
81   McKnight, Religious Foundations, 9.
82   White, Peace Among the Willows, 180.
83   Ibid., 106.
84   In *Novum Organum*, Bacon notes that inventors should hold the highest rank in society. They should be placed above political "heroes" since achievements in politics only extend to particular societies whereas inventions can benefit all societies. Furthermore, they bring "benefit without hurt or sorrow to anyone" (99).
85   White, Peace Among the Willows, 149.
86   New Atlantis, 290.\
87   Ibid., 292.
88   Ibid., 298.
89   Ibid., 300.
90   Ibid.
91   Ibid., 302.
92   Ibid.
93   He characterizes the wisdom from the Greeks as a "kind of childish stage of science . . . too ready to talk, but too weak and immature to produce anything" (*Preface to the Great Renewal*, 6).
94   Ibid.
95   For instance, Thomas Hobbes. Hobbes sees fear as a primary motivator of human action and characterizes the man's natural state as one of "continual fear and danger of violent death" (Thomas Hobbes, *Leviathan* [Oxford and New York: Oxford University Press, 1998], Chapter XIII, Paragraph 9).
96   Novum Organum, 91.
97   Ibid., 88.
98   See Rummel, *Death by Government*, 4.
99   *Novum Organum*, 101 (emphasis added).

100   As Aristotle claimed it to be over two millennia ago. See Aristotle, *Nicomachean Ethics*, trans. Martin Ostwald (Upper Saddle River, NJ: Prentice Hall Publishing, 1999), 4 (Book I, Section II).

Let There Be Light: The Newtonian Age

1   English poet Alexander Pope's famous epitaph. Quoted in I. Bernard Cohen's *Science and the Founding Fathers* (New York, NY: W.W. Norton, 1995), 121.

2   Voegelin, *English Quest for the Concrete*, 205.

3   Ibid., 206.

4   Bernard Cohen notes that Newton's use of the vernacular for the *Opticks* instead of Latin shows that he regarded it as a less systematic and scientific work, mainly because it had not been mathematized. Furthermore, he failed to include his name on the title page of the *Opticks*. xxvii. See I. Bernard Cohen, "The Case of the Missing Author: The Title Page of Newton's *Opticks* (1704)," in *Isaac Newton's Natural Philosophy*, ed. Jed Z. Buchwald and I. Bernard Cohen (Cambridge, MA: MIT Press, 2004), 15–46.

5   See Andrew Janiaks introduction. Isaac Newton, *Philosophical Writings*, ed. Andrew Janiak (Cambridge: Cambridge University Press, 2004), xii.

6   The subtitle of his *Treatise on Human Nature* of 1739 is "An Attempt to Introduce the Experimental Method of Reasoning into Moral Subjects."

7   "I feign no hypotheses." Ibid., xxiv.

8   Ibid., xxvi.

9   Such as in the Queries of the *Opticks*.

10   The mixing of metaphysics and natural philosophy was something that Bacon had adamantly opposed as well. Yet as we can see by Newton's criticisms, the practice had continued well after Bacon's time.

11   This is consistent with Kuhn's observation that those who offer a new paradigm usually work within the structure of the existing paradigm. Newton directly engaged a variety of philosophical and theological questions in spite of his insistence on purging such queries from natural philosophy. Furthermore, he spent considerable time studying alchemy and searching for the "philosopher's stone," which scientists today would consider well outside of the realm of proper

science. Many of his successors did not know about this due to the secrecy with which he carried out his studies. See Michael White's *Isaac Newton: The Last Sorcerer.*

12   Voegelin, English Quest for the Concrete, 164.

13   Ibid., 165.

14   Ibid. Drawing from Whitehead, the schism Voegelin is referring to is between those who submit to the "fallacy of misplaced concreteness" and those who can free themselves of it. Unfortunately, the former far outnumber the latter in influence.

15   Again, we must separate Newton's own life from the impact of his work. He was a highly spiritual person as is evident from his personal notes and extensive studies of alchemy. However, those studies were not made public and so the impact of his work was almost exclusively in the realm of "strict" natural science.

16   Ibid., 183.

17   Ibid.

18   Glanville argues that the *New Atlantis* served as the model for the Royal Society. Without a doubt, its founders were influenced heavily by Bacon. Jardine notes that Bacon was considered to be the "figurehead, patron saint and Father of Modern Science" by the early Royal Society. See Francis Bacon, *Novum Organum*, xviii.

19   Newton intentionally wrote the *Principia* in a way that would make it nearly impossible for a layman to understand. The original work was in Latin and his use of propositions required a mathematical background. He did not allow an English translation until the last year of his life. See White, *Isaac Newton: The Last Sorcerer*, 216.

20   For instance, his *Principia* outlined the three laws of motion as well as the theory of universal gravitation. Furthermore, he was able to reconcile Galileo's mechanics with Kepler's (ibid.).

21   For instance, he could not outline the cause of gravitation. This gets back to a primary difference between modern science and the previous framework. Questions of first and final causes were generally eschewed with the advent of modern science. Thus, Newton's work tells us *how* things work, but remains silent on *why*. Explanation is essentially relegated to description.

22   Gordon Clark, *The Philosophy of Science and Belief in God* (Unicoi, TN: Trinity Foundation, 1964), 37.

23   Ibid., 42.

24 See Richards Schwab's introduction to Jean Le Rond D'Alembert, *Preliminary Discourse to the Encyclopedia* (Chicago: University of Chicago Press, 1995), xxii.

25 Ibid., xi. Eric Voegelin calls it "the revolutionary manifesto of a new attitude towards man and society." See Eric Voegelin, *From Enlightenment to Revolution* (Durham: Duke University Press, 1975), 76.

26 Schwab, Introduction, xi.

27 Ibid., xxv.

28 Ibid., xxxi.

29 Ibid. Emphasis added.

30 Voegelin, From Enlightenment to Revolution, 74.

31 Schwab, Introduction, xxxiii.

32 Rationalism refers to the school of thought that posits the existence of absolute, certain truths that can be instinctively known. These *a priori* truths can then be used as the basis to explain phenomena. Empiricism is based on the assumption that "hard facts of experimentation, experience, and physical sensations are the essential elements from which our valid ideas are derived and which are the source of all true knowledge" (ibid., xxxii).

33 Ibid., xxxvi.

34 Ibid., xxxv.

35 Voegelin, From Enlightenment to Revolution, 165.

36 Schwab, Introduction, xxxviii.

37 Voegelin, From Enlightenment to Revolution, 82.

38 Schwab, Introduction, xl.

39 Ibid.

40 For instance, Machiavelli heavily relied on history to instruct the Prince.

41 Voegelin, From Enlightenment to Revolution, 83.

42 Ibid., 84.

43 Ibid., 79.

44 Ibid., 80.

45 D'Alembert, *Preliminary Discourse*, 26.

46 Ibid., xlvi.

47 Voegelin, From Enlightenment to Revolution, 81.

48 Ibid.

49 D'Alembert, *Preliminary Discourse*, xlviii.

50 Ibid., 43.

51 F. A. Hayek, *The Counter-Revolution of Science: Studies on the Abuse of Reason*, 2nd ed. (Indianapolis: Liberty Press, 1979), 187.

52 Ibid., 195.

53 Ibid., 195.

54 Ibid., 220.

55 Ibid., 220.

56 Ibid., 221 (emphasis added).

57 Ibid., 219. Note that the top position is to be held by a mathematician. This further reflects the influence of the *Principia* and *Discourse*.

58 Ibid.

59 Ibid., 222.

60 Ibid., 243.

61 Ibid., 249.

62 See Peyton Lyon, "Saint-Simon and the Origins of Scientism and Historicism," *The Canadian Journal of Economics and Political Science* 27.1 (1961): 62.

63 Voegelin, From Enlightenment to Revolution, 190.

64 Ibid., 191.

65 Ibid.

Auguste Comte's Scientistic Politics

1 Taken from Auguste Comte's *Preface* to *The Catechism of Positive Religion*. Cited by Voegelin, *From Enlightenment to Revolution*, 141.

2 Wernick, *Auguste Comte*, 3.

3 Ibid., 5.

4 Ibid., 20.

5 John Stuart Mill and Émile Littré both rejected the "later" Comte. In a letter to Congreve, Mill lamented, "It is M. Comte himself, who, in my judgment, has thrown ridicule on his own philosophy by the extravagance of his later writings." As quoted in Wernick, *Auguste Comte*, 23. Originally from Mary Pickering, *Auguste Comte: An Intellectual Biography*, vol. 1 (Cambridge: Cambridge University Press, 1993), 697.

6 Voegelin, From Enlightenment to Revolution, 145.

7 Auguste Comte, *Introduction to Positive Philosophy*, ed. Frederick Ferre (Indianapolis: Hackett Publishing, 1998), vii.

8   According to Kolakowski, he was part of a group that greeted Napoleon's return "too enthusiastically." See Leszek Kolakowski, *The Alienation of Reason: A History of Positivist Thought*, trans. Norbert Guterman (New York: Anchor Books, 1969), 46.

9   Comte, Introduction to Positive Philosophy, viii.

10   Ibid., viii.

11   Wernick, *Auguste Comte*, 19.

12   Comte directly imports the Baconian maxim that true knowledge is useful knowledge.

13   In *Novum Organum*, Bacon notes that inventors should hold the highest rank in society. They should be placed above political "heroes" since achievements in politics only extend to particular societies whereas inventions can benefit all societies. Furthermore, they bring "benefit without hurt or sorrow to anyone." See Francis Bacon, *Novum Organum*, 99.

14   For instance, the ninth month is named for Gutenberg.

15   Particularly Aristotle's logic and Plato's theology. Both hampered development in the "practical" sciences. For instance, Aristotle's logic is "quite divorced from practice and completely irrelevant to the active part of the sciences" (ibid., 16).

16   See preface to the *Great Renewal* (ibid., 6).

17   As Comte notes of his own development, he was "a theologian in childhood, a metaphysician in youth" (Comte, *Introduction to Positive Philosophy*, 4).

18   For instance, "empty and endless discussions between, e.g., divine right and the sovereignty of the people" (ibid., 12).

19   Bacon saw them purely as destructive hindrances. Comte recognized their futility, but thought they were *necessary* steps in the development of the mind. See Wernick, *Auguste Comte*, 40.

20   Comte, Introduction to Positive Philosophy, 12.

21   Wernick, *Auguste Comte*, 47.

22   Ibid.

23   Ibid., 48.

24   Ibid.

25   Voegelin, From Enlightenment to Revolution, 147.

26   Comte, Introduction to Positive Philosophy, 1.

27   Ibid.

28   Ibid., 1–2.

29   Ibid., 29.

30 Ibid., 2.

31 Kolakowski, *Alienation of Reason*, 51.

32 Ibid., 52.

33 Comte, Introduction to Positive Philosophy, 2.

34 Kolakowski, *Alienation of Reason*, 53.

35 Ibid., 53.

36 Comte, Introduction to Positive Philosophy, 2.

37 Kolakowski, *Alienation of Reason*, 54.

38 Ibid.

39 Ibid., 55.

40 Ibid.

41 Ibid., 56.

42 Ibid.

43 Ibid.

44 Ibid.

45 Ibid., 57.

46 Ibid., 58.

47 Ibid.

48 Comte, Introduction to Positive Philosophy, 4.

49 Ibid., 30.

50 Ibid., 4.

51 Ibid.

52 For example, "Theology and physics are so profoundly incompatible their conceptions are so radically opposed in character, that, before giving up the one in order to employ the other exclusively, the human intelligence had to make use of intermediate conceptions, which, being of a hybrid character, were eminently fitted to bring about a gradual transition" (ibid., 7).

53 Ibid., 12.

54 Ibid.

55 Wernick, *Auguste Comte*, 21. Voegelin, Hayek, Mill, and Dumas make similar claims.

56 Eric Voegelin, *Science, Politics, and Gnosticism*, ed. Ellis Sandoz (Wilmington: ISI Books, 2004), 15 (emphasis added).

57 Ibid., 17.

58 Ibid., 20.

59 F. A. Hayek, *Counter-Revolution of Science*, 222.

60 Comte, Introduction to Positive Philosophy, 17.

61 Ibid.

62   Ibid., 18.
63   Ibid., 19.
64   Ibid., 24.
65   Ibid.
66   Ibid.
67   Ibid., 25.
68   Ibid., 26.
69   Ibid., 26–27.
70   Ibid., 28.
71   Ibid.
72   Ibid.
73   Ibid.
74   Ibid., 29.
75   Ibid.
76   Voegelin, From Enlightenment to Revolution, 90.
77   Wernick, *Auguste Comte*, 34.
78   Ibid.
79   Ibid., 185.
80   "Our sole object here is to effect, for the new religion, a general ex-
     position equivalent to that which formerly taught you Catholicism.
     This second operation ought to be even easier than the former, for
     not only is your reason now mature, but the doctrine is by its na-
     ture, more intelligible as always demonstrable" (Comte, *Catechism*,
     49).
81   Wernick, *Auguste Comte*, 2.
82   Ibid.
83   Ibid.
84   Ibid.
85   Ibid., 3.
86   Ibid.
87   "It is quite as easy to see the inadequacy of every priesthood which
     aims at guiding the soul whilst taking no account of its subordina-
     tion to the body. This separation . . . must cease, once for all, by a
     wise reincorporation of medicine into the domain of the priesthood"
     (Comte, *Catechism*, 36).
88   Note the similarities to the House of Solomon in Bacon's *New At-
     lantis*.
89   Wernick, *Auguste Comte*, 3.

90   As Comte notes, women and proletariats will support the idea of a moral government because their "good sense has been left unimpaired by our vicious system of education" (ibid.).

91   Ibid., 4.

92   The Positivist Calendar includes the likes of Shakespeare, Bacon, Newton, Descartes, Anaximander, Dante, St. Paul, and Caesar.

93   Wernick, *Auguste Comte*, 4.

94   Ibid.

95   Ibid., 24.

96   Ibid.

97   "Without her I should have never been able to practically make the career or St. Paul follow on that of Aristotle, by founding the universal religion on true philosophy, after I had extracted the latter from real science" (Comte, *Catechism*, 13).

98   Ibid., 11.

99   It is important to recall that the proletariats and women make up the primary audience for Comte's project as a whole.

100   Ibid., 24.

101   Ibid., 25.

102   Comte is vague in regards to the specifics of how this transition will occur in practice. It is important to note that Comte thinks this can take place without a violent revolution.

103   Ibid.

104   Ibid., 18.

105   Ibid.

106   Ibid., 41.

107   Ibid.

108   Since such questions belong to the outmoded theological and metaphysical phases, they can no longer be asked.

109   For a pertinent contemporary example of this conception of Humanity, see James Ceaser, "The Roots of Obama Worship," *Weekly Standard*, January 25, 2010, http://www.weeklystandard.com/articles/roots-obama-worship.

110   Comte, *Catechism*, 53.

111   Ibid., 55.

112   Ibid.

113   According to Comte's instructions, we should "forget the defects of the dead in order to recall only their good qualities . . . for if our

Divinity only incorporates into herself the really worthy dead, she also takes away from each the imperfections which in all cases dimmed their objective life" (ibid., 72).

114 Moses, Homer, Aristotle, Archimedes, Caesar, St. Paul, Charlemagne, Dante, Gutenberg, Shakespeare, Descartes, Frederick II, and Bichat (a French anatomist/physiologist) are the thirteen months. See Table D in *Catechism*.

115 Ibid., 90.

116 Ibid.

117 Ibid., 91.

118 Ibid., 92.

119 Women do not receive the sacraments of destination, maturity, or retirement due to their relegation to private, moral teachers (ibid., 93).

120 Ibid.

121 Ibid., 94.

122 Ibid., 95.

123 Ibid.

124 Ibid., 67.

125 Voegelin, From Enlightenment to Revolution, 136.

126 Comte, *Catechism*, 13.

The Evolution of Scientism: Marx, Darwin, and Eugenics

1   "Historical action is to yield to their personal inventive action . . . to an organization of society specially contrived by these inventors. Future history resolves itself, in their eyes, into the propaganda and the practical carrying out of their social plans." They possess a "fanatical and superstitious belief in the miraculous effects of their social science" (Karl Marx and Frederick Engels, *Communist Manifesto*, trans. Martin Milligan [New York: Prometheus Books, 1988], 239–41).

2   Revolution can only occur when class antagonisms have fully formed and the material conditions dictate it (ibid., 240).

3   Voegelin, *From Enlightenment to Revolution*, 242.

4   "The philosophers have only interpreted the world in various ways; the point however is to change it" (from Marx's *Theses on Feuerbach*, cited in John Hallowell, *Main Currents in Modern Political Thought* [New York: Henry Holt and Co., 1950], 406).

5   Marx's program had an obvious practical appeal since workers were indeed being exploited. However, Voegelin's explanation gets at the existential crisis that allowed for the acceptance of such a program (Voegelin, *From Enlightenment to Revolution*, 255).

6   Ibid., 257.

7   Ibid., 256.

8   Voegelin suggests that Marx fundamentally misunderstood Hegel and that such a misunderstanding was not due to an intellectual error, but was deliberate. "The Idea is for Hegel, of course, not the demiurge of the 'real' in the sense in which Marx understands the term, that is in the sense of empirical reality. Rather, it is the demiurge of the 'real' only insofar as reality is the revelation of the Idea. . . . It is precisely because empirical reality and the reality of the Idea are *not* identical that the problem of the Idea arises" (ibid., 257).

9   Voegelin points to the comments made by the editors of Marx's early writings as further proof of this: "He tacitly argues from a position that is unphilosophical on principle" and the justification of this position "is simply assumed." Furthermore, "by simply referring to what in common parlance is called reality, the philosophical question concerning the nature of reality is cut off" (ibid., 258). From Siegfried Landshut's and J. P. Mayer's introduction to Karl Marx, *Der Historische Materialismus*, vol. 1 (Leipzig: Kröner, 1932), xxii.

10   Voegelin, From Enlightenment to Revolution, 260.

11   The shift can best be seen in the works of Spencer and Darwin.

12   Hallowell, Main Currents, 408.

13   As quoted by Hallowell, *Main Currents*, 408. See http://www.marxists.org/reference/archive/spirkin/works/dialectical-materialism/ch02-s03.html.

14   Ibid., 408.

15   Ibid., 408.

16   Voegelin, From Enlightenment to Revolution, 266.

17   Ibid.

18   Ibid., 267.

19   Ibid., 268. While Voegelin is right to point out the indebtedness to the Encyclopedists, I argue the impulse was started by Bacon's project.

20   Karl Marx, *Economic and Philosophic Manuscripts of 1844*, trans. Martin Milligan (New York: Prometheus Books, 1988), 110.

21 Ibid.

22 Ibid.

23 Ibid., 111.

24 Ibid.

25 Ibid.

26 Ibid. This closely mirrors Comte's suggestion that "good minds" recognize the emptiness of metaphysical questions. See Comte, *Introduction to Positive Philosophy*, 12.

27 Marx, Economic and Philosophic Manuscripts, 111.

28 Ibid.

29 Quoted by Hallowell, *Main Currents*, 402–3. Originally from Marx's "A Contribution to the Critique of Hegel's Philosophy of Right," http://www.marxists.org/archive/marx/works/1843/critique-hpr/intro.htm.

30 "Only in a communist society where every man can develop himself in any way he chooses will the true nature of man be revealed. Historical man . . . has no absolute value. Society alone can produce the true man" (ibid., 405).

31 Marx and Engels, *Communist Manifesto*, 243 (emphasis added).

32 As quoted by Hallowell, Main Currents, 428.

33 Voegelin, From Enlightenment to Revolution, 253.

34 Hallowell, *Main Currents*, 432. Originally from Engels, *Origins of the Family, Private Property and the State*, http://www.marxists.org/archive/marx/works/1884/origin-family/ch02d.htm.

35 The specifics of such a morality are not given, as they can only be determined after the revolution (ibid., 433–34).

36 Ibid., 435.

37 Ibid., 436.

38 Marx, *Economic and Philosophic Manuscripts*, 102–3 (emphasis added).

39 As Leszek Kolakowski notes, Marx's "philosophy entailed some practical consequences that would bring indescribable misery and suffering to mankind: private property and the market were to be abolished and replaced by universal and all-encompassing planning—an utterly impossible project" (Kolakowski, *Main Currents of Marxism* [New York: W.W. Norton and Company, 2005], vi).

40 Voegelin, From Enlightenment to Revolution, 242.

41 As Kolakowski notes, in spite of Marx's intentions to create a utopian end state, "the Marxist doctrine was a good blueprint for

converting human society into a giant concentration camp" (Ko-lakowski, *Main Currents of Marxism*, vi).

42 Under Stalin's regime, tens of millions died in the gulags. Whether Marx would have wanted such results is irrelevant. The actual consequences of his ideas are what we must be concerned with. For instance, Rummel estimates that 62 million Russians died at the hands of their own government in the twentieth century. See Rummel, *Death by Government*, 24.

43 For instance, with the Soviet Union and the PRC in the twentieth century.

44 And as Kolakowski points out, "the virus (of Marxism) is dormant, waiting for the next opportunity. Dreams about the perfect society belong to the enduring stock of our civilization" (*Main Currents of Marxism*, vi).

45 Darwin was heavily influenced by the likes of Herbert Spencer and Thomas Malthus. Spencer originally coined the term "survival of the fittest," which Darwin conceded to be more accurate than his term of "natural selection." Malthus's economic theory of population growth is what inspired Darwin to develop his theory of natural selection. Darwin claimed his own theory was merely "the doctrine of Malthus applied with manifold force to the whole animal and vegetable kingdoms" (West, *Darwin Day*, 106–9).

46 In fact, Darwin took great care *not* to speculate on these issues, especially in his public works. He ambiguously spoke of a Creator, but never did endorse the Judeo-Christian story of Genesis. However, in his private correspondence, Darwin expressed regret that he "truckled to public opinion, and used the Pentateuchal term of creation" when he really meant "appeared by some wholly unknown process" (ibid., 37). See also John Brooke, "Darwin and Victorian Christianity," in *Cambridge Companion to Darwin*, 206.

47 Eric Voegelin, *The New Order and Last Orientation*, ed. Jürgen Gebhardt and Thomas Hollweck, Collected Works of Eric Voegelin 25 (Columbia, MO: University of Missouri Press, 1999), 184.

48 Ibid., 185.

49 Ibid., 178. It is important to note that Voegelin does not consider phenomenalism to be a *necessary* result of the advancement of science, but also points out that its rise is hardly conceivable without it.

50 Ibid., 185.

51 I will more fully explore the historical shift to "phenomenalism" in the subsequent section on the "race idea."

52 West, *Darwin Day*, 109.

53 Ibid.

54 Ibid., 107.

55 Ibid., 34.

56 Brooke, "Darwin and Victorian Christianity," 199–200.

57 Ibid., 201.

58 Earlier in his life, Darwin was an admitted theist because he saw the "impossibility of conceiving the immense and wonderful universe, including man . . . as the result of blind chance or necessity." It is important to note Darwin generally remained ambivalent on the issue in his public statements even after the change, undoubtedly as a result of political prudence (West, *Darwin Day*, 38–39).

59 Ibid., 38.

60 It is in this work that Darwin explicitly addresses the question of human evolution. He had avoided such a discussion in *On the Origin of the Species*.

61 "The idea of a universal and beneficent Creator of the universe does not seem to arise in the mind of man, until he has been elevated by long-continued culture" (Darwin, *Descent of Man*, 682).

62 West, *Darwin Day*, 39.

63 Ibid., 40.

64 Darwin, *Descent of Man*, 86.

65 Darwin claims "all animals feel wonder, and many exhibit curiosity," dogs show a sense of humor and magnanimity, monkeys exhibit sympathy and fidelity, and suspicion is "eminently characteristic of most wild animals" (Descent *of Man*, 90–92).

66 Ibid., 110.

67 Ibid., 116.

68 Ibid.

69 "The idea of a universal and beneficent Creator of the universe does not seem to arise in the mind of man, until he has been elevated by long-continued culture" (ibid., 682).

70 West, Darwin Day, 28.

71 Ibid., 29.

72 Darwin, *Descent of Man*, 137–38.

73 West, *Darwin Day*, 31.

74 Ibid.
75 See Diane Paul, "Darwin, Social Darwinism, and Eugenics," in *Cambridge Companion to Darwin*, 234. In addition to Galton, Darwin's son Leonard was a strong supporter of eugenics as evident in his role of President of the Eugenics Society in Britain (ibid., 241). Thus while Darwin himself never fully embraced eugenics, there is ample evidence that his system led to its implementation.
76 Ibid., 235.
77 Ibid.
78 Ibid.
79 Ibid., 235–36.
80 Quoted in West, *Darwin Day*, 122. See Woodrow Wilson, *Constitutional Government in the Papers of Woodrow Wilson*, ed. Arthur Link (Princeton, NJ: Princeton University Press, 1974), 18:105.
81 West, *Darwin Day*, 122. See Woodrow Wilson, *The New Freedom*, ed. William Leuchtenburg (Englewood Cliffs, NJ: Prentice Hall, 1961), 41–42.
82 West, *Darwin Day*, 122.
83 Ibid., 123.
84 Ibid., 124.
85 This lends support to the idea of the need for scientifically trained policy makers (ibid.).
86 Ibid., 125 (emphasis added).
87 Ibid.
88 Ibid.
89 Ibid., 129 (emphasis added).
90 Ibid., 130.
91 Ibid.
92 Ibid., 132.
93 Ibid.
94 Ibid. Quote by feminist eugenicist Charlotte Gilman.
95 Ibid., 138. Justice Butler, a Roman Catholic, was the only dissenter. The case still has yet to be overturned, although *Skinner v. Oklahoma* (1942) is widely acknowledged as the turning point in quelling negative eugenics in the United States.
96 Ibid., 139. According to Kelves, it jumped fivefold from 2–4 per one hundred thousand to 20 per hundred thousand from 1920 to the end of the 1930's.

97 Paul, "Darwin," 236–37.

98 Ibid., 238.

99 Ibid.

100 Ibid., 239.

101 Ibid., 236.

102 The books were only available for a short time however and were met with harsh criticism by National Socialist reviews. See Vondung's introduction to Voegelin, *History of the Race Idea*, xi.

103 Ibid., 23.

104 Ibid., 3.

105 Ibid., 3–4.

106 Ibid., 4. Voegelin summarizes the Christian image presented by Kempis.

107 Ibid   .

108 Ibid., 7.

109 Ibid.

110 Ibid.

111 Voegelin credits Ray as perhaps the most prominent source of the race idea. This is in part because Ray's method abandoned the previous modes of classification: *differentia specifica* and *genus proximum*. His descriptive method brought "attention to the fact that the species are defined by complexes of traits" (ibid., 32–33).

112 Ibid., 93.

113 Ibid.

114 Ibid.

115 Ibid., 90.

116 Ibid., 109.

117 The monads are imperishable and divinely created. According to Leibniz, there is no substance except for monadic substance. The monad is "both body and soul at the same time" (ibid., 108).

118 Ibid., 109.

119 Ibid., 104.

120 Ibid., 112.

121 Ibid., 114.

122 Voegelin points to Linnaeus's work as an example of this view (ibid., 115).

123 Ibid., 115–16.

124 Ibid., 8.

125 Ibid., 169.

126 Carus, as cited by Voegelin (ibid., 169–70).

127 Ibid., 170.

128 Ibid., 178.

129 "In what manner the mental powers were first developed in the lower organisms, is as hopeless an enquiry as how life itself first originated. These are problems for the distant future, if they are ever to be solved by man" (Darwin, *Descent of Man*, 87).

130 Stephen Gould offers a scientific refutation of biological determinism in *The Mismeasure of Man*. Gould recognizes the dangers inherent with such a view: "we must never forget the human meanings of lives diminished by these false arguments—and we must, primarily for this reason, never flag in our resolve to expose the fallacies of science misused for alien social purposes." Gould specifically targets measures of IQ and points to the negative political consequences of such enterprises (such as the forced sterilization of "imbeciles"). Gould argues that cultural evolution is what explains changes and differences amongst societies, not genetics. Biological changes occur very slowly and simply offer potentialities (not determinants) for action. See Stephen Jay Gould, *The Mismeasure of Man*, rev. ed. (New York and London: W.W. Norton, 1996), 50 and 355.

131 Darwin repeatedly refers to natural selection as the probable cause of the persistence of these traits. Although he concedes that "conscience" is what sets man most apart from the lower animals, he nonetheless posits that certain animals possess sympathy and exhibit love (*Descent of Man*, 120–30).

132 This is what Voegelin refers to as "biological phenomenalism" (*The New Order and Last Orientation*, 184).

133 See Philip Kitchener, "Giving Darwin his Due," in *Cambridge Companion to Darwin*, 472.

134 See Alasdair MacIntyre, *After Virtue*, 2nd ed. (Notre Dame, IN: University of Notre Dame Press, 1984). MacIntyre traces the rise of managerial expertise in the twentieth century to Enlightenment principles (ibid., 79–87).

135 This is not to say that contemporary political science is radically deficient. Rather, it simply cannot answer certain types of questions; some of which are critically important to politics. Scientism represents such a case.

Towards a New Science of Man

1   *Nicomachean Ethics*, Book X, 1179A.

2   Clearly social science aims to meet these demands. But political science is primarily concerned with the question of *who* rules and the well-being of society is directly impacted by such a decision.

3   I readily acknowledge that such a task would require a full-length study. This chapter is meant to provide a preliminary outline or sketch of what a science of politics would need to include if it is to successfully address the problem of scientism.

4   MacIntyre, *After Virtue*, 82–83. Without such explanations, action becomes the result of mechanical, physiological factors.

5   Instead, it relies on material and efficient causes. It aims to describe *how* things work.

6   Both in the sense of Classical and Christian ethics. "Ethical" behavior is reduced to materialistic causes or ascribed to selfish impulses. See Hobbes, Diderot, and Darwin.

7   "To explain is on their view to invoke a law-like generalization retrospectively; to predict is to invoke a similar generalization prospectively." Social science has clearly failed to accomplish these goals. See MacIntyre, *After Virtue*, 92.

8   Ibid., 92. For a comprehensive account of how social sciences have failed to produce such results see Stanislav Andreski, *Social Sciences As Sorcery* (New York: St. Martin's Press, 1972).

9   MacIntyre, *After Virtue*, 91.

10   Ibid., 92.

11   Ibid., 92

12   Ibid., 93.

13   MacIntyre points out that a trained observer may seemingly be able to predict actions that the agent may not be able to. However, this also fails because the observer cannot "predict the impact of his future actions on my future decision-making, he cannot predict my future actions any more than his own" (ibid., 95–96).

14   This scholarship is known as "rational choice theory" within political science. For a prominent example, see Bueno De Mesquita, *War Trap* (New Haven and London: Yale Press, 1984).

15   MacIntyre, *After Virtue*, 97. A classic example of this relates to warfare. Sides often exaggerate their own relative power. If both sides had perfect information as to the capabilities of their opposition,

there would be no need to fight. Or think of a poker game. If I know what my opponent is holding, I will only call a bet if my own hand is superior.

16 Ibid., 100.

17 MacIntyre has harsh words for those that deny this fact: "omniscience excludes the making of decisions. If God knows everything that will occur, he confronts no as yet unmade decision. He has a single will. It is precisely insofar as we differ from God that unpredictability invades our lives. This way of putting the point has one particular merit: it suggests precisely what project those who seek to eliminate unpredictability from the social world or to deny it may in fact be engaging in" (ibid., 97).

18 MacIntyre points to an experiment where one hundred people were told they had to meet an unknown person in Manhattan. The only knowledge provided was that the other person had the same information. Over eighty picked the large clock in front of Grand Central Station at noon (ibid., 102).

19 Ibid., 102.

20 And perhaps of more importance for social scientists, the corollary: unpredictability does not necessarily entail inexplicability (ibid., 102).

21 For social regularities, MacIntyre gives the example of nineteenth and twentieth century Britain and Germany where class determined the type of education one would receive (ibid., 103).

22 Ibid., 104.

23 It should be noted, that some scholars have indeed treated error in a constructive way. See Eric Gartzke, "War is in the Error Term," *International Organization* 53 (1999): 567–87. Gartzke acknowledges the role that chance plays in warfare and notes that its prevalence prevents us from successfully predicting the onset of war. Uncertainty can lead to war or peace.

24 MacIntyre is referring to the rise of the modern bureaucracy and "technocratic expertise." These institutions are based on the assumption that "expert knowledge" can be obtained through social science; something that MacIntyre's argument challenges (ibid., 86).

25 And to the extent that the unpredictability is recognized, the attempt to produce law-like generalizations will be abandoned.

26 In eschewing ontological questions and abandoning "metaphysics," Enlightenment thinkers failed to recognize that they were themselves

engaging in metaphysics. In Voegelin's terms, they are merely rejecting an "idealistic metaphysic" for a "materialistic metaphysic." As
MacIntyre notes, they took part in an "unacknowledged and unrec-
ognized transition from one stance of theoretical interpretation to
another." This led him to characterize the Enlightenment as the "period *par excellence* in which most intellectuals lack self knowledge."
See MacIntyre, *After Virtue*, 81 and Eric Voegelin, *Anamnesis*, trans.
Gerhart Niemeyer (Columbia, MO: University of Missouri Press,
1978), 32. As noted previously, metaphysical questions are derided
in Bacon's work and banned outright in the systems of Comte and
Marx.

27 A view discussed at some length in Voegelin's analysis of race theory.
See Voegelin, *History of the Race Idea*.

28 Voegelin, *Anamnesis*, 32–33.

29 The constructive part of the study is especially indebted to the philosophy of Eric Voegelin and more specifically, his theory of consciousness as presented in *Anamnesis*. See also Ellis Sandoz,
"*Principia Noetica*," in *The Voegelinian Revolution*, 2nd ed. (New
Brunswick, NJ: Transaction Publishers, 2000).

30 Strauss points to this shift as one of the distinguishing features of
modernity. See Leo Strauss, *Liberalism: Ancient and Modern*
(Chicago: University of Chicago Press, 1989), 20–21.

31 "Reason is, and ought only to be the slave of the passions, and can
never pretend to any other office than to serve and obey them"
(David Hume, *A Treatise of Human Nature*, ed. L. A. Sigby-Bigge
[Oxford: Clarendon Press, 1896], Book II, Section III, Part III "Of
the influencing motives of the will").

32 To the extent the soul was thought to exist at all, discussion of it
was relegated to "unscientific" fields such as theology.

33 See Aristotle, *Nicomachean Ethics*, trans. Martin Ostwald (Upper
Saddle River, NJ: Prentice Hall Publishing, 1999). To the extent that
man leads the contemplative life, his life "would be more than
human. A man would do so not insofar as he is human, but because
there is a divine element within him" (1177b 25–30).

34 For example, see the *Phaedo* (63A–B). Socrates defines philosophy
as the practice of dying. This is because the body is viewed as a hindrance to the soul. He then proceeds to provide epideictic proofs for
the immortality of the soul.

35 I am simply referring to the new *view* of reason and the subsequent

understanding of man that results. I am not suggesting an actual change in human nature.

36 Voegelin, *Anamnesis*, 89.

37 Ibid.

38 Ibid.

39 Ibid., 90–91.

40 This is due to the anthropological principle of Plato. The polis is "man writ large." This means that the order of a society will reflect the order of the individuals that constitute it. See *Republic*, Book II (367C–369C).

41 Voegelin, *Anamnesis*, 90.

42 Ibid., 92.

43 See Plato's *Theaetetus*, trans. Joe Sachs (Newbury Port, MA: Focus Publishing, 2004), 155d.

44 Voegelin, *Anamnesis*, 93 (emphasis added).

45 See Plato's *Republic*, Book VII.

46 Voegelin, *Anamnesis*, 94.

47 Ibid., 101.

48 *Ratio* is what gives the search direction toward to the Ground (ibid., 101).

49 Ibid., 102–3.

50 Ibid., 102.

51 Voegelin asserts that those thinkers knowingly distort reality; it is not simply an intellectual error (ibid., 3).

52 Bacon, *Novum Organum*, 8.

53 And as we noted in the Chapter on Bacon's New Science, ethical and political constraints prohibit the implementation of experimentation on human beings. We cannot "vex" humans in the same way we can manipulate nature in the laboratory.

54 Ellis Sandoz, *The Voegelinian Revolution*, 2nd ed. (New Brunswick, NJ: Transaction Publishers, 2000), 148.

55 Voegelin, *Anamnesis*, 209.

56 As Aristotle readily points out in the *Nicomachean Ethics*: "[A] well-schooled man is one who searches for that degree of precision in each kind of study which the nature of the subject at hand admits." To do otherwise is "foolish" (1094b lines 23–26).

57 As quoted by Voegelin (*Anamnesis*, 211–12).

58 Ibid., 212.

59 Ibid.

60   See Marx, *Economic and Philosophic Manuscripts*, 111.

61   See Leo Strauss, *What Is Political Philosophy? And Other Studies* (Chicago: University of Chicago Press, 1988), 19–20. Stanislav Andreski makes a similar claim noting social scientists "have followed and continue to follow the intellectual fashions of the day: hurrah-patriotic in 1914, pacifist in the twenties, leftist in the thirties, celebrating the end of ideology in the fifties, youth-cultured and new leftist at the end of the sixties" (*Social Sciences as Sorcery*, 30).

62   Strauss, Political Philosophy, 20.

63   "To be a competent student of what is right and just, and of politics generally, one must first have received a proper upbringing in moral conduct" (*Nicomachean Ethics*, 1095B, lines 5–10).

64   Ibid., 1102A, lines 20–25.

65   Eric Voegelin, *The New Science of Politics: An Introduction* (Chicago and London: University of Chicago Press, 1952), 11–12.

66   Andreski points to the fact that human beings react to what is said about them as the primary reason for the impasse. "Imagine how sorry would be the plight of the natural scientist if objects of his inquiry were in a habit of reacting to what he says about them: if the substances could read or hear what the chemist writes or says about them, and were likely to jump out of their containers and burn him if they did not like what they saw on the blackboard or in his notebook" (*Social Sciences as Sorcery*, 20).

67   It should be noted that empirical statements do not have to be, and often are not, "scientific."

68   Again, I am working under the assumption that the only reality we have is *experiential* reality. I am not suggesting that normative claims be reduced to the empirical or that the distinction cannot be made. For example, Aristotle's claim that the contemplative life is the happiest is not meant as an opinion. It is a statement of fact about what is best for man and therefore is empirical. But the suggestion that man should pursue such a life is normative.

69   This is not to say that the term will not be abused if the original meaning is recovered.

70   Leo Strauss, "An Epilogue," in *Liberalism: Ancient and Modern* (Chicago: University of Chicago Press, 1995). 205

71   Hume's subtitle to *A Treatise of Human Nature* clearly indicates his intentions to follow Newton's lead: "an attempt to introduce the experimental method of reasoning into moral subjects."

72   A problem addressed fully by Alasdair MacIntyre. See *After Virtue*, 81–82.
73   Voegelin, *Anamnesis*, 177.
74   Ibid., 178.
75   Ibid., 201.
76   The "in-between." This refers to the fact that man participates in both the transcendent and immanent.
77   Sandoz, *Voegelinian Revolution*, 193–94.

Conclusion: Scientism in the Twenty-First Century and Beyond

1   Eric Voegelin, *Science, Politics, and Gnosticism*, ed. Ellis Sandoz (Wilmington: ISI Books, 2004), 17.
2   Such as the Encyclopedists and Saint-Simon.
3   Paul Feyerabend argues that science advances through insights that often do not fall within the established theoretical and methodological assumptions of a field. See Feyerabend, *Against Method*, 3rd ed. (New York and London: Verso, 1993), 2.
4   See James Shotwell, *Intelligence and Politics* (New York: The Century Co., 1921), 21 and 26–27, as quoted by Bernard Crick in *The American Science of Politics* (London: Routledge and Kegan Paul Publishing, 1959), xii.
5   It is one thing to question exactly which subjects can be investigated using a particular methodology. But methodology itself is clearly not concerned with the substance of politics.
6   See Andrew Rehfeld, "Offensive Political Theory," *Perspectives on Politics* 8 (2010): 465–86.
7   Ibid., 478.
8   See Peter Boyer, "The Covenant," *The New Yorker*, September 6, 2010, http://www.newyorker.com/reporting/2010/09/06/100906fa_fact_boyer?currentPage=all.
9   This essentially allowed federal funding for lines that were created elsewhere, but still did not allow federal funding for the creation of new lines.
10   Judge Royce Lamberth issued the injunction citing the destruction of embryos as a violation of the Dickey-Wicker Amendment of 1995 (which has been subsequently renewed by Congress). http://www.reuters.com/article/idUSTRE67M4HA20100823.

11   For instance, see Dawkins' *The God Delusion* (Boston and New York: Mariner Books, 2006) and *The Blind Watchmaker: Why the Evidence of Evolution Reveals a World without Design* (New York and London: W.W. Norton and Company, 1996) and Hitchens' *God is not Great: How Religion Poisons Everything* (New York: Twelve, 2007).

12   See Stephen Hall's article, "Darwin's Rottweiler," *Discover Magazine*, September 2005.

13   For a full discussion of the spiritual reductionism inherent in Dawkins' work see Meins Coetsier, "God the Creative Ground of Existence: A Response to Richard Dawkins' *The God Delusion*," Conference paper given at the annual meeting of the Eric Voegelin Society, Boston, MA, 2008.

14   Dawkins, *God Delusion*, 57.

15   "Our mission is to support scientific education, critical thinking and evidence-based understanding of the natural world in the quest to overcome religious fundamentalism, superstition, intolerance and human suffering" (http://richarddawkins.net/home/about).

16   The problem does not lie with the scientific findings he utilizes, but rather the extrapolations he makes as to the meaning of those findings.

17   http://www.project-reason.org/about/. See also http://samharris.org.

18   This is the interpretation offered by Voegelin. As Sandoz notes, "Faith in the reason of the Whole is the foundation . . . of all inquiry. The first principles or basic assumptions of all science rest on faith, the dimension of *Nous* that embraces *pistis*" (*Voegelinan Revolution*, 198).

19   "I have learned that, beyond death and taxes, there is at least one absolutely indisputable fact: not only does human caused global warming exist, but it is also growing more and more dangerous, and at a pace that has now made it a planetary emergency" (Al Gore, *An Inconvenient Truth* [New York: Rodale Publishing, 2006], 8).

20   The warming trend of the 1990s has not continued into the twenty-first century. The five-year mean global temperature has been stagnant for the last decade, yet the climate models touted by Gore predicted a continuous rise. A recent article in *The Economist* nicely outlines the issue: http://www.economist.com/news/science-and-technology/21574461-climate-may-be-heating-up-less-response-greenhouse-gas-emissions.

21   A good example of the politicization of science can be found in the case of Bjorn Lomborg. Lomborg's book *The Skeptical Environmentalist* and subsequent documentary, *Cool It,* both raise serious questions about the claims made by climate alarmists. Lomborg was vehemently attacked as a result and was accused of academic fraud. He was ultimately absolved of the charges. Lomborg acknowledges global warming is occurring, but notes its effects are much smaller than what Gore claims. He argues resources can be better spent on other pressing issues such as poverty and disease. See Bjorn Lomborg, *The Skeptical Environmentalist,* Reprint Edition (Cambridge: Cambridge University Press, 2001).

22   Nor do I wish to suggest that contemporary political science has not produced appreciable knowledge about politics. My main contention is that contemporary political science paints an "incomplete picture" of political reality by remaining silent on certain questions. And unless it begins to address those questions, it will continue to remain silent about the concrete political problems I have outlined in this study.

23   Questions such as Leibniz's: "Why is there something, why not nothing?" and "Why do things have to be the way they are and not different." See Sandoz, *Voegelinian Revolution,* 196–97.

24   Al Gore made similar accusations in *An Inconvenient Truth,* further illustrating the political, and partisan, nature of the "scientific" debates.

Index

Alchemy, 35
Aphorisms, 23–24, 38, 63
Apperceptive experience, 19, 136
Aristotle: ancient model of politics 9, 19; rejected by moderns, 20, 26, 28, 39, 41, 61, 64; the *bios theoretikos*, 52, 132; on facts, 124, 138; teleology, 125; on reason, 131–34; on experience, 136–37, 140, 143, 146; physics, 141.
Astronomy, 47, 55, 62, 65, 68, 69

Bell, Alexander Graham, 111
Biology, 3, 15, 47, 62, 68–69, 103, 121
Bios Theoretikos: rejection of by moderns, 15, 51, 53, 64, 92, 95, 124, 131, 145, 154; the fullness of experience, 19; the happiest life, 132
Bonaparte, Napoleon, 128
Bush, George W., 149

Carus, Ray, 119–121
Catholic Church, 12, 56, 59, 78
Chemistry, 62, 68–69, 75
China, 153
Christianity: and natural theology, 16, 104; and Baconian

science, 25, 32–35, 39, in the Newtonian Age, 45–46, 52–53, 56; rejection by Comte, 11–12, 59, 60, 63, 65, 77–78, 88–89; the New Atheists, 151; human nature, 19, 132
Citizen, 6, 19, 34, 82, 112, 143–44
Climate change, 1, 2, 152–54
Columbus, Christopher, 22, 28, 32
Common sense, 6–7, 136–37, 143
Communism, 13–15, 99–100
Condillac, Ettiene Bonnot de, 48
Consciousness: and Darwin 16, 106; and religion 96–97; noetic, 133–34, 142–43
Cyrus, 86

D'Alembert, Jean le Rond, 48, 51, 53, 77
Dawkins, Richard, 150–52
Democracy, 137–38, 144, 147
Descartes, Rene, 44, 49, 66, 75
Dewey, John, 20
Diderot, Denis, 48, 51, 53, 77
Differentia specifica, 16, 106, 121, 131

Ecole Polytechnique, 54, 60
Empiricism, 6, 49, 131, 139–40